Dr John Gribbin trained as an astrophysicist at the University of Cambridge before becoming a full-time science writer. He has worked for the science journal *Nature* and the magazine *New Scientist,* has contributed articles on science topics to *The Times*, the *Guardian* and the *Independent*, and has made several acclaimed science series for BBC Radio 4. John Gribbin has received awards for his writing in both Britain and the United States and is currently a visiting Fellow in astronomy at the University of Sussex. His many books include *In Search of Schrödinger's Cat*, *Stephen Hawking: A Life in Science* (with Michael White) and *In Search of SUSY*. John Gribbin is also the author of several science fiction works including *Innervisions*.

He is married with two sons and lives in East Sussex.

JOHN GRIBBIN

IN SEARCH OF THE
EDGE OF TIME

PENGUIN BOOKS

PENGUIN BOOKS

Published by the Penguin Group
Penguin Books Ltd, 27 Wrights Lane, London W8 5TZ, England
Penguin Putnam Inc., 375 Hudson Street, New York, New York 10014, USA
Penguin Books Australia Ltd, Ringwood, Victoria, Australia
Penguin Books Canada Ltd, 10 Alcorn Avenue, Toronto, Ontario, Canada M4V 3B2
Penguin Books (NZ) Ltd, Private Bag 102902, NSMC, Auckland, New Zealand

Penguin Books Ltd, Registered Offices: Harmondsworth, Middlesex, England

First published by Bantam Press, a division of Transworld Publishers Ltd, 1992
Published in Penguin Books 1995
Reprinted with Introduction 1998
5 7 9 10 8 6

Printed in England by Clays Ltd, St Ives plc

Contents

ACKNOWLEDGEMENTS   vii

INTRODUCTION: Why Time Travel is Possible   ix

CHAPTER ONE: Ancient History   1
Let Newton be!; On ye shoulders of Giants; Three
laws and a theory of gravity; The test of time; Across
the Solar System; Black hole pioneers; Waves and
particles: towards twentieth-century science

CHAPTER TWO: Warping Space and Time   32
From Euclid to Descartes; Beyond Euclid; Geometry
comes of age; The geometry of relativity; Einstein's
gravitational insight; The relativity of geometry;
Schwarzschild's singular solution

CHAPTER THREE: Dense Stars   64
Dwarfish companions; Degenerate stars; The white
dwarf limit; The ultimate density of matter; Inside the
neutron star; Beyond the neutron star; Puzzling
pulsars; Zwicky was right: neutron stars revealed

CHAPTER FOUR: Black Holes Abound   100
Red shifts and relativity; Radio galaxies; Quasars;
Cosmic powerhouses; X-ray stars; Celestial
powerhouses; The prime candidate; A profusion
of possibilities

**CHAPTER FIVE: Darkness at the
Edge of Time**   125
New maps of space and time; Black holes in a spin;
Singularities rule; Defeating the cosmic censor; Black
holes are cool; Exploding horizons; Centrifugal
confusion; A one-way time machine

CHAPTER SIX: Hyperspace Connections   162
The Einstein connection; Charging through
hyperspace; Bridging the universes; The blue-shift
block; Parting the blue sheet; Journey into hyperspace;
Wormhole engineering; Making antigravity; The
string-driven spaceship: a practicable proposition?

**CHAPTER SEVEN: Two Ways to Build
a Time Machine**   198
Paradoxes and possibilities; Time loops, and other
twists; Tachyonic time travellers; Gödel's universe;
Tipler's time machine; Wormholes and time travel;
Paradoctoring the paradoxes

CHAPTER EIGHT: Cosmic Connections   230
Blowing bubbles; Einstein's vanishing constant; An
oscillating universe?; The black hole bounce

GLOSSARY   246

BIBLIOGRAPHY   254

INDEX   259

Acknowledgements

The seed of the idea for this book was planted by Igor Novikov, of Moscow State University, when he gave a seminar on the topic of time machines at the University of Sussex in 1989. At that time, I was deeply immersed in writing a book about global warming (*Hothouse Earth*, published by Bantam Press in the UK and Grove Weidenfeld in the US in 1990), but Novikov's talk, and a discussion with him afterwards, revived my long-standing interest in the more bizarre implications of the general theory of relativity. During the months that followed, in between my writing on climate change I sought out expert advice on the latest developments in the mathematical investigation of spacetime wormholes and time travel, and also refreshed my understanding of the long history of the investigation of black holes. At last, after the World Climate Conference that was held in Geneva in November 1990, I felt that there was little more I could add to the debate about the greenhouse effect, at least for the time being, and I was able temporarily to put aside concern about the fate of the planet, and to write the book you now hold. As light relief from concern about global environmental issues, it was fun to write, and helped to keep me sane. I hope you enjoy it as well, even though there is no way in which anything described in these pages can have any conceivable practical use for anyone at present living on Planet Earth (unless, of course, one of you is a time traveller who got here through use of one of the devices described in Chapter Seven).

In addition to Novikov, the experts who helped to remind me that science is fun (which was why I got into it in the first place) and who provided help in the form of discussions, copies of their own scientific papers, and/or constructive criticism of some of my own misconceptions, are (in no particular order of merit): Ian Redmount, Clifford Will and Matt Visser, of Washington University, St Louis, Missouri; Michael Morris and Kip Thorne, of CalTech; Felicity Mellor and Ian Moss, of the University of Newcastle upon Tyne; Paul Davies, of the University of Adelaide, South Australia; Werner Israel, of the University of Alberta, Edmonton; Frank Tipler, of Tulane University, New Orleans; Roger Penrose, of the University of Oxford; Stephen Hawking, of the University of Cambridge; Sidney Coleman, of Harvard University; and (by no means least) Sir William McCrea, of the University of Sussex, who had no direct input to the present book but who, more than a quarter of a century ago, was the person who first opened the door to the world of the general theory for me.

I have also had the good fortune to be in the hands of an editor, John Michel of Harmony Books, who understands physics and whose suggestions led to distinct improvements in the presentation of my ideas – for which I am grateful, in spite of his stubborn refusal to add an 'l' to the spelling of his surname.

They share the credit for any accurate insights into science that you may glean from these pages; the misconceptions that remain in the book are, of course, entirely my responsibility.

John Gribbin
June 1991

INTRODUCTION

Why Time Travel is Possible

Since the first edition of this book appeared, physicists have found the law of nature which prevents time travel paradoxes, and thereby permits time travel. It turns out to be the same law that makes sure light travels in straight lines, and which underpins the most straightforward version of quantum theory, developed half a century ago by Richard Feynman.

As I explain in this book, relativists have been trying to come to terms with time travel for the past decade, since Kip Thorne and his colleagues at Caltech discovered – much to their surprise – that there is nothing in the laws of physics (specifically, the general theory of relativity) to forbid it. Among several different ways in which the laws allow a time machine to exist, the one that has been most intensively studied mathematically is the 'wormhole' (see Chapter Six). This is like a tunnel through space and time, connecting different regions of the Universe – different spaces *and* different times. The two 'mouths' of the wormhole could be occupying the same space, but separated in time, so that it could literally be used as a time tunnel.

Building such a device would be very difficult – it would involve manipulating black holes, each with many times the mass of our Sun. But they could conceivably occur naturally, either on this scale or on a microscopic scale.

The worry for physicists is that this raises the possibility of paradoxes, familiar to science fiction fans. For example, a time traveller could go back in time and accidentally (or even deliberately) cause the death of her granny, so that neither the time traveller's mother nor herself was ever born. People are hard to describe mathematically, but the equivalent paradox in the relativists' calculations involves a billiard ball that goes into one mouth of a wormhole, emerges in the past from the other mouth, and collides with its other self on the way into the first mouth, so that it is knocked out of the way and never enters the time tunnel at all (see Chapter Seven). But, of course, there are many possible 'self-consistent' journeys through the tunnel, in which the two versions of the billiard ball never disturb one another.

If time travel really is possible – and after several years' intensive study all the evidence says that it is – there must, it seems, be a law of nature to prevent such paradoxes arising, while permitting the self-consistent journeys through time. Igor Novikov, who holds joint posts at the P. N. Lebedev Institute, in Moscow, and at NORDITA (the Nordic Institute for Theoretical Physics), in Copenhagen, first pointed out the need for a 'principle of self-consistency' of this kind in 1989.* Now, working with a large group of colleagues in Denmark, Canada, Russia and Switzerland, he has found the physical basis for this principle.

It involves something known as the principle of least action (or principle of minimal action), and has been known, in one form or another, since the early seventeenth century. It describes the trajectories of things, such as the path of a light ray from A to B, or the flight of a ball tossed through an upper-storey window. And, it now seems, the trajectory of a billiard ball through a time tunnel.

Action, in this sense, is a measure both of the energy involved in traversing the path and the time taken. For light (which is always a special case), this boils down to time alone, so that the principle of least action becomes the principle of least time, which is why light travels in straight lines.

You can see how the principle works when light from a source in air enters a block of glass, where it travels at a slower speed than in air (Figure i). In order to get from the source A outside the glass to a point B inside the glass in the shortest possible time, the light has to travel in one straight line up to the edge of the glass, then turn through a certain angle and travel in another straight line (at the slower speed) on to point B. Travelling by any other route would take longer.

The action is a property of the whole path, and somehow the light (or 'nature') always knows how to choose the cheapest or simplest

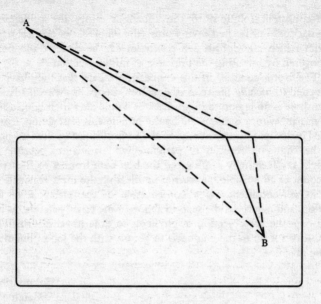

Figure i *Light travels faster through air than through glass. So the quickest journey from A to B that is partly through air and partly through glass is not the (dotted) straight line from A to B, but there is a unique 'path of least time' made up of two straight lines. This is a special case of the principle of least action at work. The dotted lines to the right show an example of a path that takes longer than the path of least time (solid lines).* ★

path to its goal. In a similar fashion, the principle of least action can be used to describe the entire curved path of the ball thrown through a window, once the time taken for the journey is specified. Although the ball can be thrown at different speeds on different trajectories (higher and slower, or flatter and faster) and still go through the window, only trajectories which satisfy the principle of least action are possible.

Novikov and his colleagues have applied the same principle to the 'trajectories' of billiard balls around time loops, both with and without the kind of 'self collision' that leads to paradoxes. In a mathematical *tour de force*, they have shown that in both cases only

★ Taken from Gribbin and Gribbin, *Richard Feynman: A Life in Science*, Viking, London, 1997.

self-consistent solutions to the equations satisfy the principle of least action – or in their own words, 'the whole set of classical trajectories which are globally self-consistent can be directly and simply recovered by imposing the principle of minimal action'.*

The word 'classical' in this connection means that they have not yet tried to include the rules of quantum theory in their calculations. But there is no reason to think that this would alter their conclusions. Feynman, who was entranced by the principle of least action, formulated quantum physics entirely on the basis of it, using what is known as the 'sum over histories' or 'path integral' formulation, because, like a light ray seemingly sniffing out the best path from A to B, it takes account of all possible trajectories in selecting the most efficient.

So self-consistency is a consequence of the principle of least action, and nature can be seen to abhor a time travel paradox. Which removes the last objection of physicists to time travel in principle – and leaves it up to the engineers to get on with the job of building a time machine.

John Gribbin
March 1998

* NORDITA preprint number 95/49A.

CHAPTER ONE

Ancient History

In which we meet Isaac Newton, and learn how a hopeless cattleherd invented a theory of gravity and indulged in academic rows. We bid farewell to the fifth force, find out how to measure the speed of light, and discover how an eighteenth-century parson used gravity to trap light in black holes

Black holes are products of gravity. Modern science began with Isaac Newton, who, among other things, developed the first scientific theory of gravity, a little over three hundred years ago. For the first time, scientists were able, using Newton's laws, to explain the motion of heavenly bodies in terms of the same principles that applied to the behaviour of objects on Earth. In the famous analogy, both the fall of an apple from a tree and the orbit of the Moon about the Earth could be explained by the same set of equations. Newton's description of gravity, of course, was later incorporated within Albert Einstein's general theory of relativity, and black holes are generally regarded, rightly, as essentially relativistic objects. But it is some indication of the power of Newton's own theory that less than a hundred years after the publication of his epic volume the *Philosophiae Naturalis Principia Mathematica*, generally regarded as the most important single book ever published in physics, and equally generally referred to simply as the *Principia*, Newtonian gravitational theory had already been used to describe what we would now call black holes. Indeed, the surprise is that Newton himself, who investigated the nature of light as well as gravity, did not realize that his equations suggested the existence of dark stars in the Universe, objects from which light could not escape because gravity would overwhelm it.

Let Newton be!

Newton was born in Woolsthorpe, Lincolnshire, on Christmas day, 1642, the same year that Galileo Galilei died (curiously, more than two centuries later Albert Einstein was born in the same year, 1879, that the greatest nineteenth-century physicist, James Clerk Maxwell, died). He was a small, sickly baby, who surprised his mother (his father, also called Isaac, had died three months before young Isaac was born) by surviving his birthday; he went on surviving for another eighty-four years. His contribution to establishing science and the scientific method as providing the best description of the material world, and the awe in which he was held by his contemporaries, were neatly encapsulated early in the eighteenth century by the poet Alexander Pope, with his famous couplet:

> Nature and Nature's laws lay hid in night:
> God said, *Let Newton be!* and all was light.

But, as we shall see, it wasn't quite that simple.

Before Newton was two, his mother remarried and moved to a nearby village, leaving him in the care of his grandmother for nine years, until the death of his stepfather. The trauma of this separation almost certainly explains Newton's strange behaviour as an adult, including his secretiveness about his work, his obsessive anxiety about how it would be received when it was published, and the violent, irrational way in which he responded to any criticism by his peers. After his stepfather died, however, Isaac and his mother were reunited, and she planned initially for him to take over the management of the family farm. He proved hopeless at this, preferring to read books rather than to herd cattle, so he was sent back to school in Grantham, and then (with the aid of an uncle who had a connection with Trinity College in Cambridge) on to university. He arrived in Cambridge in 1661, a little older than most of the other new undergraduates because of his interrupted schooling.

Newton's notebooks show that even as an undergraduate he kept abreast of new ideas, including those of Galileo and the French philosopher René Descartes. These marked the beginning of the new view of the Universe as an intricate machine, an idea which had yet to penetrate, officially, the great universities of Europe. But he kept all this to himself, while he also made a thorough study of the distinctly old-fashioned official curriculum (based on the ancient teaching of Aristotle) and obtained his bachelor's degree in 1665, a satisfactory, but seemingly not brilliant, student in the eyes of his teachers. The same year, plague broke out in London, the university was closed as a result, and Newton went home to Lincolnshire, where he stayed for the best part of two years, until normal academic life resumed.

It was during those two years that Newton derived the inverse square law of gravity – perhaps stimulated by watching the fall of an apple. In order to do this, he invented a new mathematical technique, differential calculus, which made the calculations more straightforward. And, as if this were not enough, he also began his investigation of the nature of light, discovering and naming the spectrum, the rainbow pattern of colours that is produced when white light passes through a prism. None of this made any impact on the scientific world at the time, because Newton didn't tell anybody what he was up to. When the university reopened in 1667, he was elected to a fellowship at Trinity College, and by 1669 he had developed some of his mathematical ideas to the point where they were circulated to the *cognoscenti*. By now, at least some of the professors in Cambridge were beginning to take notice of his ability, and when Isaac Barrow resigned from the post of Lucasian Professor of Mathematics in 1669 (in order to devote more time to divinity), he recommended that Newton should be his successor. Newton became Lucasian Professor at the age of twenty-six – a secure position for life (if he wanted it to be), with no tutoring responsibilities but the requirement to give one course of lectures each year. The present Lucasian Professor, incidentally, is Stephen Hawking.

Between 1670 and 1672, Newton used these lectures to develop his ideas on light into the form which later became the first part of his epic treatise *Opticks*. But this was not published until 1704, as a result of one of the most protracted personality clashes of even Newton's tempestuous career. The problems began when Newton started to communicate his new ideas through the Royal Society, an organization which had been founded only in 1660, but which was already established as the leading channel of scientific communication in Britain. The row, with Robert Hooke, also led to the most famous remark made by Newton – and one which, recent research suggests, has been misinterpreted for three hundred years.

On ye shoulders of Giants

The Royal Society first learned of Newton as a result of his interest in light – not his new theory of how colours are formed, but his practical skill in inventing the first telescope to use a mirror, instead of a lens system, to focus light. The design is still widely in use and known to this day as a Newtonian reflector. The learned gentlemen of the Society liked the telescope so much, when they saw it in 1671, that in 1672 Newton was elected a Fellow of the Society. Pleased in his turn by this recognition, in that same year Newton presented a paper on light and colours to the Society. Robert Hooke, who was the first 'curator of experiments' at the Royal Society, and is remembered today for Hooke's Law of elasticity, was regarded at the time (especially by himself) as the Society's (if not the world's) expert on optics, and he responded to Newton's paper with a critique couched in condescending terms that would surely have annoyed any young researcher. But Newton had never been able, and never learned, to cope with criticism of any kind, and was driven to rage by Hooke's comments. Within a year of becoming a Fellow of the Royal Society and first attempting to offer his ideas through the normal channels of communication, he had retreated back into the safety of his Cambridge base, keeping his thoughts to himself and avoiding the usual scientific toing and froing of the time.

But early in 1675, during a visit to London, Newton heard Hooke, as he thought, saying that he now accepted Newton's theory of colours. He was sufficiently encouraged by this to offer the Society a second paper on light, which included a description of the way coloured rings of light (now known as Newton's rings) are produced when a lens is separated from a flat sheet of glass by a thin film of air. Hooke immediately complained, both privately and publicly, that most of the ideas presented to the Society by Newton in 1675 were not original at all, but had simply been stolen from his (Hooke's) work. In ensuing correspondence with the secretary of the Society, Newton denied this, and made the counter-claim that, in any case, Hooke's work was essentially derived from that of René Descartes.

Things were brewing up for an epic row when, seemingly under pressure from the Society, Hooke wrote Newton a letter which could be interpreted as conciliatory (if the reader were charitable) but in which he still managed to repeat all his allegations and to imply that, at best, Newton had merely tidied up some loose ends. It was this letter that provoked Newton's famous remark to the effect that if he had seen further than other men, it was because he stood on the shoulders of giants.

This remark has traditionally been interpreted as indicating Newton's modesty, and his recognition that earlier scientists such as Johannes Kepler, Galileo and Descartes had laid the foundations for his laws of motion and his great work on gravity – which is odd, because in 1675 Newton hadn't made his ideas about gravity and motion public. The charge of modesty does not, in any case, seem one which would stick to such a prickly, even arrogant, character as Newton, although it is easy to see how the story might appeal to later generations. So where did the remark come from?

In 1987, as part of the celebrations marking the tercentenary of the publication of the *Principia*, Cambridge University organized a week-long meeting at which eminent scientists from around the world brought the story of gravity up to date. At that meeting, John Faulkner, a British researcher now based at the Lick Obser-vatory in California, presented his persuasive new interpretation of what Newton meant by that remark, based on Faulkner's probing into the documents related to the feud with Hooke. Newton was certainly not being modest, but arrogant when he made that statement, said Faulkner; and he was certainly not referring to Kepler and Galileo, or his work on gravity, but, indeed, to his work on light.

In fact, similar references to the giants of the past were common in Newton's day, and were generally used to express indebtedness to the ancients, especially the Greeks. Seventeenth-century scient-ists in general (perhaps even Newton himself) seem to have thought that they were doing no more than rediscovering laws known in much more detail to the ancients. Newton's choice of words, in a letter to Hooke dated 5 February 1675, seems to have been particularly careful, bearing in mind their previous disagreements and the fact that Hooke himself had a distinctly unprepossessing personal appearance.

Quoting from seventeenth-century contemporaries of Newton and Hooke, including Hooke's friends, Faulkner created a picture of Hooke resembling nothing so much as William Shakespeare's caricature of Richard III – distinctly twisted, and even dwarfish. Even taking some of this with a pinch of salt, there is no doubt that Hooke was a little man.

In this context, says Faulkner, the sentences in Newton's letter leading up to the remark about giants show that remark in a quite different light. Remember that this was, after all, not a hurried note despatched to a friend but a letter written at the behest of the Royal Society in order to resolve publicly an embarrassing quarrel between two of its Fellows; Newton certainly chose his words carefully to achieve that objective, but in view of his previous and subsequent behaviour it seems more than likely that, as Faulkner suggests, he took equal care with the hidden sub-text. Here are the relevant sentences, with Faulkner's interpretation of Newton's intended meaning:

'What Des-Cartes did was a good step.' (Interpretation: he did it before you did.) 'You have added much in several ways, & especially in taking ye colours of thin plates into philosophical consideration.' (Interpretation: all you did was follow where Descartes led.) 'If I have seen further it is by standing on ye shoulders of Giants.' (Interpretation, taking particular notice of Newton's careful use of the capital 'G'; *my* research owes nothing to anybody *except* the ancients, least of all to a little runt like you.)

Taking the exchange of letters at face value, they achieved the Society's objective of pouring public oil on troubled waters and restoring respectability to the dealings between its Fellows. But the upshot was that Newton retreated back even further into his shell following this encounter. He waited patiently until Hooke died, in 1703, before publishing his *Opticks* in 1704, when he could safely have the last word. And it was only through the intervention of his friend Edmund Halley, of comet fame, that he was pushed into publishing his greatest work, the *Principia*, in 1687, twelve years after the second row with Hooke. By then, the core of the work was more than twenty years old.

Three laws and a theory of gravity

Newton's *Principia* contains the heart of what is known as classical mechanics – the three laws of motion, and a theory of gravity. The shoulders on which he could indeed be said to have stood, metaphorically, in developing these ideas were those of Johannes Kepler, a German astronomer who published the first two laws of planetary motion that now bear his name in 1609. Kepler himself developed those laws using tables of planetary positions painstakingly compiled by the Dane Tycho Brahe, who had settled in Prague, where Kepler became his assistant, and who died in 1601.

Kepler's first and second laws state that the orbits of the planets around the Sun are ellipses, not circles, and that a line joining a planet to the Sun traces out equal areas in equal times, wherever the planet is in its orbit (Figure 1.1). In other words, each planet moves

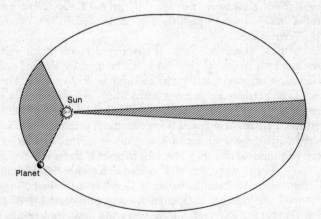

Figure 1.1 *A planet in an elliptical orbit around the Sun moves faster when it is closer to the Sun, so that the area swept out by the moving planet in a chosen time is always the same.*

faster when it is closest to the Sun, tracing out a short, fat triangle at one end of the ellipse, and slower when it is furthest from the Sun, tracing out a long, thin triangle at the other end of the ellipse. A third law, published several years later, relates the orbital period of each planet to the diameter of its orbit by a mathematical formula.

All this was intriguing and puzzling for seventeenth-century scientists, who searched unsuccessfully for an underlying explanation of Kepler's laws. Newton himself was not a complete scientific recluse, even in the late 1670s and early 1680s, and he engaged in correspondence with Hooke concerning the behaviour of objects falling under the influence of gravity – correspondence which, almost inevitably, later led Hooke to accuse Newton of having stolen the idea of an inverse square law from him. We can only imagine the surprise Halley must have felt in August 1684 when he visited Newton in Cambridge and, on mentioning that he was interested in the problem of orbital motion, was told by Newton that he had solved that puzzle years ago. Whatever his surprise, Halley kept his wits about him. He persuaded Newton that this was such a significant discovery that it simply had to be published, and just three months later Newton sent Halley a short paper on the subject. But this was not enough. Once Newton decided to publish his ideas, he began revising and rewriting this short paper until it grew into his great book, published (largely at Halley's expense) in Latin in 1687 – it wasn't published in English until 1729, two years after Newton died.

Even then, however, Newton harboured some of his secrets. Although his personal papers show that he actually worked out his famous law of gravity using the calculus he had invented, in the *Principia* he presented it in a reworked form, using essentially geometrical techniques that would, literally, have been intelligible to Aristotle. Perhaps he was just being secretive; perhaps, recalling his undergraduate days, he had a low opinion of the intellect of his peers, and thought that they might be happier with the old-fashioned approach. Either way, this was to lead to another bitter wrangle, this time with Wilhelm Leibnitz, a German mathematician who developed the calculus independently and published his work in 1684. Today, there is no doubt that Newton came up with the idea first, nor is there any doubt that Leibnitz knew nothing of this when he came up with it, and they are given equal credit for the invention. At the time, it caused yet another of Newton's bitter feuds.

For my present story, though, what matters is what was in the *Principia*, not what Newton chose to leave out. Before Newton, scientists accepted the Aristotelean idea that the 'natural' state of an object is to be at rest, and to move only when a force is applied to it. Newton realized that this only seems to be the case because we live on the surface of a planet, where things are held down by gravity. His first law says that every object (scientists usually use the term 'body') continues in a state of rest *or of uniform motion in a straight line* unless a force acts upon it. His second law states that the acceleration of a body (the rate at which its velocity changes, which means that either its speed *or* direction of motion changes) is proportional to the force acting upon it. And his third law says that whenever a force is applied to an object there is an equal and opposite reaction force – if I push a pencil across my desk, for example, or press down on the desk-top itself, I can feel the reaction to the force I am applying, a force pushing back on my fingertip. And although you might think, according to the second law, that the force of gravity should be making us accelerate towards the centre of the Earth, as long as we are standing on solid ground the force of our weight pressing down is countered by an equal and opposite reaction force pushing up. The two forces cancel out, so there is no acceleration – unless you fall over, or jump out of the window. If that happens, what hurts when you hit the ground is *not* the force of gravity, but the reaction force of the ground, cancelling out the force of gravity and stopping your motion.

Using his own three laws and Kepler's laws, Newton explained the motion of the planets around the Sun, and of its moons around Jupiter, as the result of a force of gravity which is proportional to one over the square of the distance between the Sun and a planet, or between Jupiter and a moon. This is the famous inverse square law. So, for example, when a planet is closest to the Sun the force it feels is stronger, and it moves more quickly as a result. What's more, Newton said that this was not a special law that applied only to the orbits of planets around the Sun, but a universal law which described the effect of gravity on *everything* in the Universe. The neatest example of this is one that Newton himself gave.

I already assumed that gravity applies in the same way to objects falling to the ground as to planets in their orbits when I referred to someone falling to the ground. Familiarity with the notion makes it seem obvious to us today, but in Newton's day it was a new and revolutionary idea. I also mentioned that the force of gravity acting on an object falling to the Earth acts as if all the mass of the Earth were concentrated in a point at the centre of the planet – the distance that comes into the inverse square law is actually the distance between the centres of the two bodies involved, whether they be the Sun and a planet, the Earth and a falling human being, or whatever. In fact, Newton proved this; it is a key feature of his theory of gravity, and the hardest part to prove mathematically, especially the way he did it in the *Principia*, without using calculus.

Newton also knew that the acceleration caused by gravity near the surface of the Earth will make any body (an apple, for example) fall through a distance of sixteen feet in the first second of its fall (I'll use old-fashioned feet and inches in this example, because those are the units Newton used). The Moon is sixty times further away from the centre of the Earth than the distance from the centre of the Earth to the surface, and according to Newton's first law it would 'like' to travel in a straight line at a constant speed – that is, at constant velocity. Even if the speed stays the same, any deviation from that straight line must be due to a force, deflecting the Moon. According to the inverse square law, the force of the Earth's gravity acting at the distance of the Moon should be less than the force at the surface of the Earth by a factor of sixty squared, which is 3,600. So in one second the Earth's gravity should make the Moon shift sideways by a distance given by dividing sixteen feet by 3,600. This works out at a little bit more than one twentieth of an inch. For an object travelling at the speed of the Moon, at the distance of the Moon from the Earth, a sideways nudge of exactly this size, every second, is exactly enough to make it travel in a closed orbit around the Earth, completing one circuit every month.

Newton really had explained the fall of an apple and the motion of the Moon with one set of laws. In doing so, he removed the mystery from the behaviour of heavenly bodies, and opened the eyes of scientists to the fact that the behaviour of the stars and planets – the behaviour of the whole Universe – might be explained using the same laws of physics that are derived from studies carried out in laboratories on Earth. Today, many physicists believe that they may soon be able to find a single set of equations that will describe all of the particles and forces of nature in one package – a Theory of Everything, or TOE. If they ever achieve this goal, it will be the culmination of more than three hundred years of progress along the path first trodden by Newton, and Newtonian physics will, in a sense, have come to an end. But, as we shall see, that would not necessarily mean that everything in the Universe was thoroughly understood.

Even in Newton's day, it was clear that another layer of understanding must underpin his famous inverse square law. To be sure, Newton showed that the gravitational force exerted by the Earth, or the Sun, or, indeed, anything else, falls away as one over the square of the distance from the centre of the object. But why should it be an inverse *square* law? Why not a force that falls off as one over the distance, or as one over the distance cubed, or some other law entirely? Newton did not know, and he seems not to have been concerned *why* gravity should obey an inverse square, rather than some other law. In another famous comment, addressing this very point, he wrote *Non fingo hypotheses*, which means 'I do not frame hypotheses'. He was content to explain *how* gravity did its work, without worrying about *why* it did so; and it took more than two hundred years, following the publication of the *Principia*, for anyone to solve the puzzle. Whatever the 'why', though, there is certainly no doubt that gravity does follow Newton's inverse square law.

The test of time

The force of gravity exerted by a body is proportional, in fact, both to one over the square of the distance from its centre and to its mass. A more massive object exerts a stronger gravitational pull. The gravitational pull exerted by the Earth at its surface is what we call weight. Each gram of matter in a body at the surface of the Earth is pulled by the Earth with the same strength, so the more matter there is in a body the more weight it has – the heavier it is. We say that the force exerted by the Earth on a mass of one gram at its surface is one gram weight – on Earth, a *mass* of one gram *weighs* one gram, which is a logical definition for people who live on the surface of the Earth to use. But things are not so simple for spacefarers. If we move a body with a particular mass – say, one kilogram – from the Earth to the Moon, it would still have the same mass, in this case a thousand grams. But because the Moon has less mass than the Earth, each gram would be pulled by the Moon less strongly than it was pulled by the Earth when it was back home on the surface of the Earth. So it would have less weight – in fact, a one-kilo mass on the surface of the Moon would only weigh about a sixth of a kilo.

This prediction of Newton's theory has, of course, been tested directly – people have been to the Moon, and observed the difference in weight. Nobody seriously expected otherwise – apart from anything else, the calculation of the trajectory that the spacecraft followed used Newton's laws, and they would never even have got to the Moon if those laws had been incorrect. But still, it's nice to know that Newton was right at this 'gut reaction' level. In fact, during the 1980s there was a flurry of excitement among scientists, which leaked over into the media, as a result of suggestions that at a much more subtle level Newton might, after all, have been wrong – that there might be a tiny deviation from the inverse square law of gravity at distances of a few tens of metres, even though it works perfectly for calculating planetary orbits and spacecraft trajectories. The excitement proved unfounded, but because of all the fuss Newton's gravitational law has now been tested more accurately than ever before, and come out with flying colours.

One way of looking at all this is in terms of the constant of proportionality that comes into the law. If the gravitational force acting on each gram is proportional to the mass of the Earth and to one over the square of the distance from the centre of the Earth, this is the same as saying that the force is *equal* to a constant (called G) multiplied by the mass of the Earth and by one over the square of the distance to its centre. Part of Newton's powerful insight was that even when we are dealing with different masses and different distances (such as the mass of the Sun acting on the Earth at a distance of 150 million kilometres), the constant, G, is the same. Intriguingly, though, Newton himself never used the term 'gravitational constant' in the *Principia*. He had no need to, because all of his calculations, such as the one comparing the fall of an apple to the orbit of the Moon, can be carried out in terms of ratios, and in such cases the constant cancels out of the equation.

In the 1730s, the French physicist Pierre Bouguer investigated the density of the Earth by measuring the deflection of a plumb line near a mountain, and in principle these measurements can be used to calculate G. But the first really accurate measurements that provide a measure of the gravitational constant were only made in the 1790s, more than a hundred years after the publication of the *Principia*, by Henry Cavendish, a British physicist who seems to have been even more reticent than Newton about publishing his results.

Cavendish was an eccentric recluse who published very little during his lifetime (he died in 1810 at the age of seventy-eight). He could afford this indulgence because he inherited considerable wealth from his uncle; he was the son of Lord Charles Cavendish, himself a Fellow of the Royal Society, and grandson of (on his father's side) the Duke of Devonshire and (on his mother's side) the Duke of Kent. When he died, he left more than a million pounds, an extraordinary fortune in those days; part of the family fortune was used in the 1870s by a later Duke of Devonshire (the seventh, himself a talented mathematician) to establish the Cavendish Laboratory in Cambridge, named after Henry and to this day a centre of scientific excellence. Long after Henry Cavendish died, studies of his papers showed that he had anticipated a great deal of the work on electricity, in particular, later carried out by others, including Ohm's law of resistance. Cavendish's electrical researches were eventually edited by James Clerk Maxwell, the first Cavendish Professor of Physics and head of the new Cavendish Laboratory, and published in 1879. His gravity measurements, however, were published during his lifetime, in 1798. Like Bouguer's earlier studies, they were designed to measure the mass and density of the Earth, and Cavendish's papers made no mention of G. But from Newton's law of gravity, once you know the mass of the Earth (and its radius) you can find G simply by measuring the weight of an object at the surface of the planet; so Cavendish's experiments are regarded as the first accurate determination of the gravitational constant. What's more, the method he used to make those measurements (which was actually suggested by a certain John Michell, of whom more shortly) has become the classic laboratory test of its kind, still used, with minor modifications.

The apparatus, called a torsion balance, consists of a delicate rod, suspended at its centre by a thread, with a small weight (Cavendish used spheres made of lead) attached to each end. Two large masses (bigger spheres made of lead) were placed at an angle to the rod, so that the gravitational pull of the large masses acting on the small masses would twist the rod. Cavendish measured the angle the rod was twisted by, using a system of mirrors, and worked out the force exerted by the large spheres on the small spheres. The force of the Earth's gravity acting on the small spheres (their weight) turned out to be 500 million times greater than the sideways force exerted by the large spheres, but Cavendish managed to measure the tiny deflections involved and, by comparing these with the weights of the spheres, work out the mass of the Earth. His measurements show that the Earth contains 6×10^{24} (a 6 followed by 24 zeroes) kilos of matter, and its density is five and a half times that of water, which is what he wanted to know. Just as in the comparison between the apple and the Moon, the gravitational constant cancels out of this part of the calculations. But the equations also show, when you turn them around a little, that the value of G is 6.7×10^{-8}, in the cgs system of units.

It was another hundred years before the accuracy of Cavendish's torsion balance experiments was improved on, and it was, indeed, only in the 1890s that scientists actually began to quote a value for G, and to treat it as a fundamental constant of nature in the way that we regard it today. There is now a huge amount of evidence from laboratory experiments that it really is a constant, the same for all masses, whatever they are made of. And there is a huge amount of evidence from astronomical measurements that G has the same value everywhere, while both laboratory measurements of falling weights and astronomical studies confirm that the law of gravity is indeed an inverse square law. But in all the time since Newton, very few experiments had been carried out to measure the strength of gravity over distances from a few tens of metres to a few hundred metres. This was partly because such measurements are difficult, and partly because there seemed little point – if Newton's law works on smaller and larger scales than this, it seems natural to expect that it operates in the middle range of distances, as well. But this was the loophole that led to the flurry of excitement I mentioned earlier.

The suggestion that something might be wrong with Newton's law of gravity stemmed primarily from measurements of the force of gravity made down mine shafts. In effect, this means weighing an object very carefully at different depths, and seeing how the weight changes as you go deeper below the surface. If the Earth were a perfectly uniform sphere, at each depth below the surface the force of attraction towards the centre of the Earth would be exactly the same as if all the matter below that level were concentrated at the centre. The gravitational influence of the shell of material above the depth at which measurements are being made has no overall effect – the pull in one direction from the smaller amount of mass directly overhead and nearby is exactly counterbalanced by the pull in the opposite direction from the larger amount of mass in the same shell at a greater distance on the opposite side of the Earth.

In the real world, measurements of the force of gravity inside the Earth (and even on the Earth's surface) have to take account of geology. Different kinds of rock have different densities, and will pull on the measuring apparatus more or less strongly as a result. But in the early 1980s it did seem, particularly from a series of measurements carried out down a mine in Australia, that over a range of a hundred metres or so these experiments showed a deviation from Newton's law, as if G were about 1 per cent less than the value determined by laboratory experiments and by studies of planetary motions. Measurements made by lowering instruments down boreholes (both in the rock and in ice sheets), and by carrying them up tall towers (weighing things at different heights above the ground), seemed for a time to confirm these strange anomalies, and physicists began to talk excitedly of a 'fifth force' that acted in the opposite way to gravity (antigravity) but had a range of only a few tens of metres.* When some of the tower measurements seemed to show an extra force of attraction also at work, as well as gravity, they even started talking about the 'sixth force'. But it was all pie in the sky; Newton was right, after all. It turned out that all of the 'non-Newtonian' effects that had been claimed could, in fact, be explained by good old Newtonian gravity, if proper allowance were made for the geological distribution of the rocks and ore deposits around the sites where the measurements had been made. The original Australian 'evidence' for 'non-Newtonian' gravity, for example, turned out to be due to the straightforward Newtonian gravitational force produced by a series of ridges about three kilometres away from the mine – which gives you some idea of just how sensitive these measurements are.

Of course, it is virtually impossible to prove that there is no fifth force at work. But what physicists can do is set limits on how strong it can be without showing up in their experiments. By 1990, those limits had been narrowed down to a requirement that any fifth force must be at least 100,000 times weaker than gravity in the range of 1 metre to 1,000 metres. But the mythical fifth force can be said to have served a useful scientific purpose. It was only because some people thought that a fifth force might exist that in the second half of the 1980s physicists were encouraged to make the painstaking measurements which set these tight limits. The result is that the constancy of G and the accuracy of the inverse square law are now known better than ever to hold on all length scales from tabletop experiments to the motion of the stars and planets. We know, better than even Newton himself knew, that Newton's law of gravity really is universal.

But although he lacked the experimental proof that his law of

---

* 'Fifth' force because they already know of four others – gravity itself, electromagnetism, and the so-called strong and weak nuclear forces which only operate on subatomic levels.

gravity was universal in this sense, Newton did, of course, believe that the law must apply everywhere, and for all bodies. Since his other great achievement involved the study of light, and an explanation of its behaviour in terms of tiny particles, or corpuscles, streaming out from a light source and being reflected by mirrors or refracted by prisms and lenses, this makes it all the more remarkable that he seems never to have wondered how gravity would affect light. The first publication of any insight into that mystery had to wait almost a hundred years after the publication of the *Principia*, until the same John Michell who dreamed up the torsion balance experiment came up with the notion of dark stars.

Across the Solar System

The key to this idea, apart from Newton's law of gravity, was the measurement of the finite speed of light. And one of the biggest surprises, to most people coming across these ideas for the first time, is that the speed of light was actually measured reasonably accurately *before* Newton published the *Principia*.

The calculation was made in the 1670s by a Dane, Ole Rømer, who had been born in 1644 and was then working at the Paris Observatory. Among other things, Rømer studied the behaviour of the moons of Jupiter, which were especially interesting to astronomers of the time because they represented a miniature version of the Solar System described by Copernicus and Kepler, with the huge planet Jupiter orbited by a set of moons in much the same way that the Sun is orbited by the planets. One of Rømer's more senior colleagues in Paris was the Italian-born astronomer Giovanni Cassini, who had come to France in 1669, at the age of forty-four, to take charge of the new observatory, and who became a French citizen in 1673 (changing his given name to 'Jean' at the same time). Cassini was a skilful observer, using the latest instruments in the new observatory. In 1675 he discovered the gap which divides the ring system of Saturn into two parts and is still known as 'Cassini's division'; his more important work, though, also included studies of the behaviour of the satellites of Jupiter, and he carried out the first reasonably accurate measurement of the distance from the Earth to the Sun. It was these two sets of information that Rømer put together to work out the speed of light.

One of the most obvious, and interesting, features of the behaviour of the moons of Jupiter is the way in which they are regularly eclipsed as they move into and out of the shadow of Jupiter itself. Even before he left Italy, Cassini had worked out a table of eclipses (rather like a bus timetable) for the four main satellites of Jupiter – Io, Europa, Ganymede and Callisto – that had been discovered by Galileo, using the first astronomical telescope, in January 1610. Using Kepler's laws to describe the motion of these moons, Cassini was able to forecast when they would be eclipsed. But Rømer found that sometimes the eclipses were a little early, and sometimes a little late, compared with the data in Cassini's tables. Concentrating on the behaviour of the innermost large moon of Jupiter, Io, he found a regular pattern to this behaviour. The gap between eclipses was shorter than it ought to be when two successive eclipses were observed while the Earth was moving towards its closest to Jupiter (with both planets on the same side of the Sun), and longer when two eclipses were observed while the Earth was moving towards its furthest from Jupiter (on the opposite side of the Sun).

Even without knowing why this should be so, Rømer could make predictions on the basis of the pattern he had discovered. In September 1679, he predicted that the eclipse of Io by Jupiter due on 9 November would be 10 minutes later than the standard orbital calculations suggested. The prediction was borne out, and Rømer stunned his colleagues by explaining the delay as being due to the finite time that it took for light to cross space from Io to the Earth.

In the months leading up to that eclipse, the Earth had been moving in its orbit away from Jupiter. So when the previous eclipse had occurred, the light signalling that the eclipse had happened had not had so far to travel to reach the Earth. The November eclipse really had happened at the calculated time, said Rømer, but by then the Earth was so much further away from Jupiter that the light had taken an extra 10 minutes to cross space to the telescopes of the Paris Observatory.

This was where Cassini's most important work, the investigation of the size of the Solar System, came in. In 1672, Cassini had carefully observed the position of Mars against the background stars from Paris, while his colleague Jean Richer made similar observations from Cayenne, on the north-eastern coast of South America. From these measurements, they were able to work out the geometry of an enormously tall, thin triangle, with a baseline stretching nearly 10,000 kilometres* from Paris to Cayenne, and with Mars at its tip. This gave Cassini an estimate of the distance to Mars, from which he could work out the sizes of the orbits of other planets, including the Earth, using Kepler's laws and the time it takes each planet to travel around its orbit.

Cassini's estimate of the distance from the Earth to the Sun (now known as the Astronomical Unit, or AU) was 138 million km, by far the most accurate estimate made up to then – Tycho had come up with a figure of eight million km, and Kepler himself made the distance about 24 million km, while modern measurements indicate that the AU is actually 149,597,910 km. Using Cassini's estimate for the distance across the Earth's orbit, and therefore for the extra distance light from the November 1679 eclipse had to travel before reaching his telescope, Rømer calculated that the speed of light must be, in modern units, some 225,000 km per second. In fact, using Rømer's own calculation but with the modern estimate for the size of the Earth's orbit, the figure would have been 298,000 km per second; the established value for the speed of light today is 299,792 km per second, so tantalizingly close to a nice round number that some people have seriously suggested redefining the length of a metre so that the speed of light is precisely 300,000 km per second.

Whatever the actual number that comes out of the calculation, however, the real sensation surrounding Rømer's work was the claim that the speed of light was indeed finite, and that light signals did not travel instantaneously across the void of space. This was such a dramatic claim, indeed, that many scientists of the time refused to accept it. General acceptance that the speed of light is finite came only after Rømer's death. He died in 1710, but it wasn't until the middle of the 1720s that an English astronomer, James Bradley, worked out the speed of light by a different technique, leaving no room for any more doubts.

Bradley (who became the third British Astronomer Royal after Halley died in 1642) found that when he studied the bright star Gamma Draconis in September he had to tilt his telescope at a slightly different angle to that required to get a clear image of the same star in March. It was as if the star moved slightly across the sky and back again during the course of a year – and the effect, which he called aberration, occurs for all stars. But Bradley realized

* From now on, I'll stick to modern units.

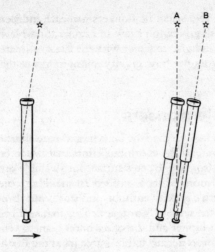

Figure 1.2 *Because the Earth is moving, a telescope has to be tilted to let light from a star travel down the tube of the telescope. A star that is really at position A seems to be at position B. The shift is in the opposite direction when the Earth is moving the opposite way, so this makes the apparent positions of the stars change during the course of a year, as the Earth orbits the Sun. The effect is called aberration, and it can be used to measure the speed of light.*

that this is actually due to the motion of the Earth through space. The extra tilt of the telescope is needed to allow for the fact that during the tiny fraction of a second it takes light to travel down the telescope tube, the telescope has been shifted sideways by the Earth's motion (Figure 1.2). Bradley measured the angular displacement of the star caused by this effect, which is a little over 20 seconds of arc; this displacement is just over 1 per cent of the angular size of the Moon as seen from Earth. By measuring this tiny displacement of starlight, he found that the speed of light must be 308,300 km per second, sufficiently close to Rømer's figure to persuade scientists in the eighteenth century that the speed of light is indeed finite, and very close to the modern value.* By the end of

---

* Bradley could, in fact, have used his value for the speed of light and the information about the eclipses of Io to calculate the diameter of the Earth's orbit and the size of the Astronomical Unit, turning Rømer's famous calculation on its head. But nobody at the time seems to have thought of this novel way to work out distances across the Solar System.

that century, two scientific thinkers had each independently hit on the idea of using Newton's law of gravity, and Newton's concept of the nature of light, together with the latest estimates of the speed of light, to work out how gravity might influence the behaviour of light.

Black hole pioneers

Anyone who has watched the launch of a space shuttle, even if only on TV, is aware of the enormous effort that has to be expended to lift an object from the Earth's surface into a stable orbit around the Earth. Even more effort is needed to make an object break free from the Earth's gravity entirely, and travel out through the Solar System, like the famous Voyager probes that sent back spectacular pictures from Jupiter and the other outer planets. The best way to measure the effort needed to break free from the Earth in this way is in terms of the speed with which the escaping object moves. For any source of gravitation (which means any object in the Universe), there is a critical speed which has to be reached before an object leaving its surface vertically can escape. It is called the escape velocity. If you could magically make the Earth more dense, so that it contained more mass but stayed the same size, the escape velocity would increase. But although very large objects like the Sun and Jupiter contain much more mass than the Earth, this is spread out across a larger volume, so that the surface of the Sun or Jupiter is much further from its centre than the surface of the Earth is from its own centre. Remember that gravity falls off as one over the square of the distance from the centre of a body – so this dilutes the strength of gravity and compensates for at least some of the extra mass. Therefore the escape velocity from the surface of a more massive (but larger) planet is not simply proportionately larger than the escape velocity from the surface of the Earth, but also depends on the density of the planet.

A rocket, like the space shuttle, builds up speed gradually as it uses fuel during takeoff. But we could achieve the same effect if we had a cannon powerful enough to fire cannonballs upwards at escape velocity. If we did this from the surface of the Earth, and fired the cannonballs straight up, they would have to leave the muzzle of the gun with a speed of 40,000 km per hour (11 km per second) in order to escape from the Earth's gravitational grip. Any moving with less initial speed will slow down, come to a halt, and then fall back to Earth; any moving faster than escape velocity will be slowed down, but never brought to a halt, and will continue moving out across space until they come under the gravitational influence of another massive object. The escape velocity from the Moon is just 8,570 km per hour, and the escape velocity from Jupiter is nearly 220,000 km per hour (just over 60 km per second).

In each case, the escape velocity is the speed with which cannonballs would have to be fired vertically upward to escape from the planet. What if we could mount our hypothetical cannon on the surface of the Sun? There, the escape velocity would be more than two million km per hour – a speed which sounds truly impressive, until you realize that it is only 624 km per second, which may be nearly 57 times the escape velocity from the surface of the Earth, but is still only 0.2 per cent of the speed of light. So light has no difficulty escaping from the surface of the Sun.

In the eighteenth century, scientists thought of light as being made up of corpuscles, in the way Newton had described, which could be visualized as very much like tiny cannonballs spat out from a glowing object. It was natural to guess that these corpuscles would be affected by gravity in just the same way as any other object, and it was straightforward to work out the escape velocity from the Earth, and to make a reasonable guess at the escape velocity from the Sun, assuming it had the same density as the Earth. But suppose there were objects in the Universe even bigger than the Sun. Suppose, indeed, that there were some stars so big that the escape velocity from the surface *exceeded* the speed of light. They would be invisible! This outrageous notion was suggested by John Michell in 1783, and caused a major stir amongst the sober ranks of the Fellows of the Royal Society.

Michell had been born in 1724, and was seven years younger than his friend Henry Cavendish. At the height of his scientific career, he was regarded as second only to Cavendish in the ranks of English scientists, and today he is still known as the father of the science of seismology. He studied at the University of Cambridge, graduating in 1752, and his interest in earthquakes was stimulated by the disastrous seismic shock that struck Lisbon in 1755. Michell established that the damage had actually been caused by an earthquake centred underneath the Atlantic Ocean. He was appointed Wood-wardian Professor of Geology in Cambridge in 1762, a year after becoming a Bachelor of Divinity. In 1764 he became rector of the parish of Thornhill, in Yorkshire, and some books give the impression that the Reverend John Michell was simply a country parson and a dilettante, amateur scientist, when in fact his scientific reputation was well founded before he entered the Church; he had already been elected a Fellow of the Royal Society (in 1760) before becoming a Bachelor of Divinity.

Michell made many contributions to astronomy, including the first realistic estimate of the distances to the stars, and the suggestion that some pairs of stars seen on the night sky are not simply chance alignments of two objects at quite different distances along the line of sight, but really are 'binary stars', in orbit around each other. And, as I have mentioned, he suggested the torsion balance method of determining the gravitational force, although he died, in 1793, before the measurements were actually carried out. In spite of all this, Michell's name was almost forgotten in the nineteenth and twentieth centuries, to such an extent that in spite of a recent rehabilitation of his reputation, his brief entry in the *Encyclopedia Britannica*, for example, does not even mention what now seems his most prescient and dramatic piece of work.

The first mention of dark stars was made in a paper by Michell which was read to the Royal Society by Cavendish on 27 November 1783, and was published the following year.* This was an impressively detailed discussion of ways to work out the properties of stars, including their distances, sizes and masses, by measuring the gravitational effect on light emitted from their surfaces. Everything was based on the supposition that 'the particles of light' are 'attracted in the same manner as all other bodies with which we are acquainted', because gravitation is, said Michell, 'as far as we know, or have any reason to believe, an universal law of nature.' Among the many other detailed arguments in Michell's long-forgotten, but now famous, paper, he pointed out that:

> If there should really exist in nature any bodies whose density is not less than that of the sun, and whose diameters are more than 500 times the diameter of the sun, since their light could not arrive at us . . . we could have no information from sight; yet, if any other luminiferous bodies should happen to revolve about them we might still perhaps from the motions of these revolving bodies infer the existence of the central ones with some degree of probability, as this might afford a clue to some of the apparent irregularities of the revolving bodies, which would not be easily

explicable on any other hypothesis; but as the consequences of such a supposition are very obvious, and the consideration of them somewhat beside my present purpose, I shall not prosecute them any further.

What Michell had realized, in modern language, is that a sphere 500 times bigger than the Sun (about as big across as the Solar System out to Jupiter) and with the same density as the Sun would have a surface escape velocity greater than the speed of light. Although the idea stirred excited debate in London, as papers still in the files of the Royal Society show, it seems not to have spread outside England, for in 1796 Pierre Laplace, seemingly in complete ignorance of Michell's proposal, put forward essentially the same idea in his semi-popular book *Exposition du système du monde*.

Bearing in mind the political changes France was going through at that time, it is not so surprising that Laplace had not been keeping up to date with his reading of the *Philosophical Transactions of the Royal Society*; he was too busy surviving, something he proved to be very good indeed at. He had been born in 1749, in Normandy, the son of a local farmer who was also a magistrate, and may have been involved in the cider business. Pierre attended a college run by the Benedictines until he was sixteen, then studied at Caen University for two years before moving on to Paris in 1768, without bothering to complete his degree. There, he impressed the mathematician Jean d'Alembert with his ability, and became a professor at the Ecole Militaire. In 1773 he was elected a member of the French Academy of Sciences, and he worked for the government both before and after the Revolution, serving on the Commission of Weights and Measures which introduced the metric system and as a senator under Napoleon (echoing, in a way, Newton's public career as Master of the Royal Mint). In 1814, sensing which way the political wind was blowing, Laplace voted for the restoration of the monarchy; his reward was to be made a marquis by Louis XVIII, and he remained active in public life, now as a supporter of the Bourbons, until his death in March 1827 (exactly one hundred years, to the month, after the death of Newton). The miracle is that with all that going on he managed to do any science at all; in fact, he was extraordinarily prolific, in some ways a French counterpart to Newton, and among other things he dotted the 'i's and crossed the 't's of Newton's own application of gravitational theory to the Solar System.

Newton himself had been baffled by one feature of the behaviour of the planets. One planet on its own, orbiting the Sun, would indeed move in a perfect ellipse in obedience to Kepler's laws, under the influence of the inverse square law of gravity. But with two or more planets, the extra gravitational forces would tug them out of their Keplerian orbits. Newton feared that these effects might lead to instability, eventually tumbling the planets out of their orbits, and sending them either crashing into the Sun or drifting away into space. He had no scientific answer to the problem, but suggested that the hand of God might be required from time to time, to put the planets back in their proper orbits before such perturbations became too large.

In the mid-1780s, however, Laplace proved that these perturbations are actually self-correcting. Using the particular example of Jupiter and Saturn, the two largest planets in the Solar System, with the strongest gravitational pulls, he found that although one orbit might contract gradually for many years, in due course it would expand again, producing an oscillation around the pure Keplerian orbit with a period of 929 years. This was one of the foundations of what is possibly the most famous remark every made by Laplace. When his work on celestial mechanics, as these studies are called, was published in book form, Napoleon commented to Laplace that he had noticed that there was no mention of God in the book. Laplace replied, 'I have no need of that hypothesis.'

Laplace's version of the dark star hypothesis – he called them 'des corps obscurs', which translates as 'invisible bodies', and obviously thought their existence more likely than that of God – was essentially the same as Michell's. One minor difference was that Laplace described his dark stars in terms of objects with the density of the Earth, which is rather greater than the density of the Sun, and therefore calculated a diameter 250 times, rather than 500 times, that of the Sun. He suggested that there might exist

> in heavenly space invisible bodies as large, and perhaps in as great number, as the stars. A luminous star of the same density as the earth, and whose diameter was two hundred and fifty times greater than that of the sun, would not, because of its attraction, allow any of its rays to arrive at us; it is therefore possible that the largest luminous bodies of the universe may, through this cause, be invisible.

The account of dark stars appeared in the first edition of the *Exposition*, published in 1796, and the second, published in 1799. In 1801, the German astronomer Johann von Soldner calculated how a ray of light passing near a star would be bent by the influence of Newtonian gravity, and even speculated that the stars that make up the Milky Way might be orbiting a very massive, central 'corps obscur' of the kind proposed by Laplace (but he decided that they probably were not, because he thought, incorrectly, that if they were, the sideways motion that would result ought to have been detected). In the edition of the *Exposition* published in 1808, however, and in all later editions, every reference to dark stars was deleted. Why did Laplace abandon the idea? It may well have been because the Newtonian image of corpuscles of light, streaming through space like tiny cannonballs, no longer seemed accurate, following the work by Thomas Young in England and Augustin Fresnel in France that revealed light to behave like a wave.

Waves and particles: towards twentieth-century science

Newton had been able to explain the properties of light in terms of corpuscles. In particular, the evidence that light travels in straight lines told him that it could not be a wave – anyone who has dropped a pebble into a pond and watched the waves spreading out will know that waves do not travel in straight lines. One obvious example of just how straight the lines that light travels in are can be seen by looking at a shadow; there is no light getting round the corners behind the illuminated object to wash out the shadow. Even light that has travelled all the way across space from the Sun and past the Moon during a solar eclipse still produces a clear shadow with nice sharp edges on the surface of the Earth.

What Young and Fresnel found, however, is that light *does* behave like a wave, but at a much more subtle level than these examples can show. The key experiment actually involves passing light through two very thin slits in a screen, and letting the light that gets through the slits fall on to another screen. The pattern of bright and dark stripes produced on the second screen shows that light has spread out, as a wave, from each of the two slits, and that the light waves have interfered with each other, just like the interference produced by the two sets of waves when you drop two pebbles into a still pond at the same time. The reason the interference effects don't show up more obviously is that light has a wavelength of about three thousandths of a centimetre, tiny compared with even the smallest ripples on a pond. With delicate enough measuring equipment, though, it is even possible to see how light does leak around the edges of objects to partly fill in their shadow – provided the object being illuminated has a very sharp edge, like a razor blade.

So whereas in the 1720s, when Newton died, almost all scientists had been convinced that light was made up of a stream of particles, by the 1820s, when Laplace died, almost all scientists were convinced that light was produced by a form of waves. Later in the nineteenth century, James Clerk Maxwell discovered the equations that describe how these waves propagate through space in the form of electromagnetic variations. A varying electric component creates the varying magnetic component, which in turn creates the varying electric component, and so the wave moves along. When Maxwell developed his equations (which explain the behaviour of radio waves, which are also known to be produced by electromagnetic effects), he found that the equations automatically fix the speed of the electromagnetic wave, and that the speed built into the equations is the speed of light. There could be no more convincing evidence that both light and radio are waves that obey Maxwell's equations.

So it came as something of a shock to physicists when, early in the twentieth century, Albert Einstein pointed out that some of the properties of light could still only be explained in terms of a stream of particles. In particular, in 1905 he explained the way in which a beam of light can knock electrons out of a metal surface (the photoelectric effect) as due to the impact of successive particles of light, exactly equivalent to Newton's corpuscles. There is no simple way in which a pure wave, even an electromagnetic wave, can do this. Einstein's work prompted a re-examination of the nature of light, which led to the shattering conclusion that it could only be explained if it behaved *both* as a wave *and* as a stream of particles, now known as photons. In 1921, Einstein received the Nobel Prize in physics for this work. So, by the 1920s, just two hundred years after the death of Newton, physicists were convinced that both Newton and Young were right, and that light is both particle and wave.

This wave-particle duality has implications that go far beyond the study of light. It is a cornerstone of the quantum theory, which describes the behaviour of the world at the subatomic level, and experiments carried out during the 1920s showed that electrons, entities that had previously been thought of as particles, also had a wave nature to their character. It is now clear that this wave-particle duality applies to all entities, although it only becomes important on the molecular and subatomic scale. Even so, as we shall see later, quantum effects can influence the behaviour of black holes.

The discovery that light has a dual nature did not, however, destroy the validity of Maxwell's equations. Light *is* still a wave, as well as being a particle; in particular, even if for some purposes, such as explaining the photoelectric effect, it is convenient to think of light as being composed of photons, these photons must still travel at the speed of light required by Maxwell's equations. But these equations make no allowance for light to be slowed down by the influence of gravity as it leaves a star – not even by the immense surface gravity of one of Michell's dark stars. In other words, force does *not* make photons accelerate. Einstein realized that Maxwell's equations and Newton's laws of motion were incompatible, and he developed the special theory of relativity (also published in 1905) to resolve the dilemma.

The cornerstone of the special theory is the fact that the speed of light in space is always the same, wherever it is measured from and however fast (and in whatever direction) the person doing the measuring is moving. It is this theory that says that all observers moving at constant velocity relative to each other are each equally entitled to regard themselves as at rest, with every other observer in motion. It explains how moving clocks run slow (because time itself is slowed by motion), moving rulers shrink, and moving objects gain mass, compared to such a stationary observer. It also tells us that energy and mass are interchangeable, and, most important of all in the present context, it says that no object can ever travel faster than light. In other words, if dark stars like those envisaged by Michell and Laplace do exist, *nothing at all* can ever escape from them. All of these effects, it is important to realize, have now been tested and measured directly in experiments involving fast-moving particles. Although the special theory of relativity runs counter to our commonsense, this is because the relativistic effects become important only at sizeable fractions of the speed of light, and our commonsense developed in a world where things do not move at such speeds.

But Einstein appreciated that he still did not have a complete theory of the Universe, such as Newton presented in the *Principia*, because his special theory deals only with constant velocities, not accelerations. To describe acceleration, and gravity, he developed the general theory of relativity, published in its complete form in 1916. This is the theory that deals with the bending of spacetime, and which explains (indeed, *requires*) the existence of black holes in the Universe. It shows that even though light always travels at the same speed (denoted by c), objects just the size envisaged by Michell and Laplace will indeed trap light and be dark.

The first hints that this might be the case, after the general theory was published, echoed the speculation of von Soldner more than a century before. Einstein's new theory predicted a deflection of light from the stars as it passed close by the Sun, but a different amount of deflection from that expected in the old Newtonian theory. Nobody had ever looked for this deflection, partly because by the time it became possible to carry out the necessary tests everyone knew that light was a wave, and so would not be affected in the way von Soldner had suggested at all. But both waves and particles (or dual wave-particles) would be deflected, according to Einstein's theory, by space bent by the presence of the Sun's mass to act like a lens. But how can you see stars in the daytime? The only way to test this prediction was to wait for a total eclipse of the Sun, when it is possible to photograph the stars which lie in the direction of (but far beyond) the Sun. If the Sun is bending space to make it act like a lens, the apparent positions of these stars will be slightly shifted. Comparing these photographs with ones taken six months later, when the Sun is on the opposite side of the sky as viewed from Earth, and the same stars are now visible at night, you can see if any of the images have been shifted. Fortuitously, there was a total solar eclipse in 1919; the photographs were taken and compared, and proved Einstein's theory to be correct (Figure 1.3). The story made headline news, suggesting (not entirely accurately) that Newton's theory had been overturned, and Einstein became a household name.

The discovery of light bending initiated a little flurry of what turned out to be premature speculation by a few theorists, unwittingly echoing the forgotten speculations of Michell and Laplace in a more modern form. Now, however, for the first time those speculations considered what would happen to the escape velocity of a star like the Sun if it were squeezed into a smaller sphere, so

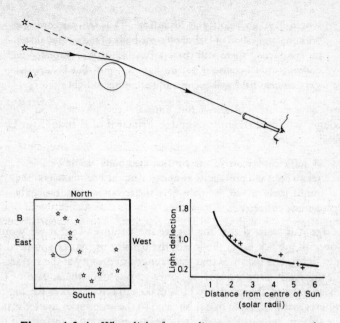

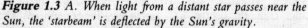

Figure 1.3 *A. When light from a distant star passes near the Sun, the 'starbeam' is deflected by the Sun's gravity.*
B. During the solar eclipse of 1919, a team led by Arthur Eddington measured the light-bending effect for several stars. The amount by which the light was deflected for the different stars (crosses on the graph) exactly matched the predictions of Einstein's general theory of relativity (solid line).

that the distance from the surface to the centre got smaller and gravity at the surface increased, even though the mass stayed the same. A researcher from University College, Galway, commented in 1920★ that:

> We may remark, though perhaps the assumption is very violent, that if the mass of the Sun were concentrated in a sphere of diameter 1.47 kilometres, the index of refraction near it would be infinitely great, and we should have a very powerful condensing lens, too powerful indeed, for light emitted by the sun itself

★ A. Anderson, *Philosophical Magazine*, Volume 39, pp. 626–8. Quoted by Werner Israel in his contribution to *300 Years of Gravitation*, edited by Stephen Hawking and Werner Israel; see the Bibliography for details.

would have no velocity at its surface. Thus if, in accordance with the suggestion of Helmholtz, the body of the sun should go on contracting there will come a time when it is shrouded in darkness, not because it has no light to emit, but because its gravitational field will become impermeable to light.

Just a year later, in the same journal★ the physicist Sir Oliver Lodge, who had recently retired as Principal of Birmingham University, wrote:

A sufficiently massive and concentrated body would be able to retain light and prevent its escaping. And the body need not be a single mass or sun, it might be a stellar system of exceedingly porous character . . .

Lodge had realized that the bigger the volume of what we would now call a black hole, the lower the density of matter you need to trap light. The reason is that the strength of gravity at the surface of the sphere is proportional not only to one over the square of the distance from the centre (which makes gravity weaker for bigger spheres with the same mass) but also to the amount of matter inside the sphere, which, for a chosen density, increases as the *cube* of the distance from the centre. For bigger and bigger spheres with the same density, the overall effect is that the strength of gravity, and the escape velocity, at the surface increase exactly in line with the increase in radius. Double the radius, and you double the escape velocity. You can make a black hole out of anything at all, with any density you like, if you have enough of it to fill a big enough sphere. Lodge realized that a star system like our Milky Way Galaxy but bigger, containing thousands or millions of billions of stars spread over a sphere with a radius of thousands of light years, could have an overall escape velocity bigger than that of light, even though the stars, planets and people within the system were in no way unusual. We could be living inside a black hole, and not even notice. But he also realized that if atoms could be squeezed together so tightly that their nuclei touched one another, it would be possible to make a black hole without using much more mass than there is in the Sun.

All these ideas were about half a century ahead of their time, and nothing came of them in the 1920s. Science simply wasn't ready to take the concept of dark stars – let alone dark galaxies – seriously. But while the physicists were more concerned with other problems – sorting out the new quantum theory and using Einstein's mass-energy relationship to explain how stars can be kept hot for so long – the mathematical foundations of the study of black holes as warps in spacetime were already being laid. Indeed, the foundations were already being laid in the first half of the nineteenth century, by Karl Gauss, Nikolai Lobachevsky and János Bolyai.

Having zipped all the way through scientific history from Newton to Einstein, and dipped a toe into the deep waters of twentieth-century physics, it is time to draw back a little to look at this nineteenth-century mathematics and how the idea of non-Euclidean geometry, developed more fully in the second half of that century by Bernhard Riemann, directly influenced Einstein's work on the general theory of relativity.

CHAPTER TWO

Warping Space and Time

Problems with parallel lines, and how a fly provided a lazy philosopher with the key to studying curves. Bending geometry to curve space and close the Universe; putting geometry into relativity. How drum majorettes explain the theory of relativity. The rubber sheet universe, and the rediscovery of black holes

To a physicist, ancient history begins with Isaac Newton, in the seventeenth century. The history of geometry is both longer and shorter. Longer if you go back to the time of the ancient Greeks, two thousand years and more ago, when the basic principles that we all learned in school – the angles of a triangle add up to 180°, parallel lines never meet, and so on – were laid down; but shorter if you are interested in the kind of geometry that describes warped spacetime and which explains *why* gravity obeys an inverse square law. Even mathematicians realized the possibility of this non-Euclidean geometry only in the nineteenth century, and it wasn't until the twentieth century that physicists made any practical application of these ideas to the Universe we live in.

From Euclid to Descartes

Euclid, who lived around 300 BC, got his name attached to the 'standard' form of geometry not because he invented (or discovered) it all, but because he wrote it all down in a treatise of thirteen books called the *Elements*. He lived in Alexandria, and may have studied at Plato's Academy in Athens, though probably not until after Plato's death in 340 BC. He was not a great mathematician in his own right (certainly no Archimedes), but he lived at the end of the great period of Greek mathematical investigations, and he wrote everything out clearly, using logical arguments to prove various geometrical properties (such as the fact that the angles of a triangle add up to 180°) starting out from a few basic axioms – definitions of what we mean by a 'point', or a 'straight line', and so on. The *Elements* was translated first into Arabic, then into Latin, surviving and remaining the basis for mathematics for more than two thousand years.

But one of Euclid's basic postulates, concerning parallel lines, always proved troublesome. This is known as the parallel postulate, and it says that if you have a given straight line and a point which is not on the straight line, there is one, and only one, straight line which can be drawn through the point and parallel to the first line. Although the notion seems like commonsense, and trial and error with a ruler, pencil and paper will convince most people that it must be true, it actually turns out to be impossible to prove the parallel postulate using the other axioms of Euclidean geometry. In 1733 the Italian mathematician Girolamo Saccheri showed that the parallel postulate must hold as long as there is at least one triangle that does have angles which add up to precisely 180°; he actually considered the possibility of triangles in which this is not the case, but thought, wrongly, that he had proved that they could not exist. So Saccheri missed discovering the possibilities of non-Euclidean geometry. Euclidean geometry itself, however, had been transformed in the seventeenth century by the work of René Descartes.

Descartes was born in 1596, the son of a councillor in the parliament of Brittany, in France. He was a sickly child, and got into the habit of lying in bed, thinking. He was educated at a Jesuit college and then studied law at the University of Poitiers, graduating in 1616. But instead of settling down to a quiet life as an academic or a lawyer, he spent most of the next dozen years or so serving in various European armies, putting his mathematical talents to use as a military engineer. It was while the army of the Duke of Bavaria, in which he was then serving, was in winter quarters on the banks of the River Danube, on 10 November 1619, that Descartes, lying snug in his bed, came up with his revolutionary insight into geometry. We know the exact date and the circumstances of the discovery because Descartes himself later recounted them in his epic book *A discourse on the Method of rightly conducting the Reason and seeking Truth in the Sciences*, published in 1637 and usually referred to simply as *The Method*.

By then, Descartes had settled in the Netherlands, after leaving military service in 1629. He should, we can now appreciate, have stayed in Holland; but in 1649 he could not resist an invitation from Queen Christina of Sweden to become a member of her court in Stockholm, to found an academy of sciences, and to teach her philosophy. To his horror, on arriving in Stockholm Descartes, now in his fifties, found that instead of being able to lie in bed in the mornings, his duties included visiting the Queen at 5 a.m. each day for her personal tuition. In the Swedish winter, he soon caught a chill, which developed into pneumonia and (with the aid of the enthusiastic bleeding that doctors used to treat the illness) finished him off. He died in February 1650, a few weeks short of his fifty-fourth birthday.

The geometrical insight that Descartes had described in *The Method* remained, however, along with many other works that place him in the first rank of philosophers and scientific thinkers, not just of his day but of *any* day. He was in many ways the first modern thinker, refusing to accept anything that could not be proved beyond reasonable doubt, concluding that even the workings of the human body could be explained by basic mechanical and scientific principles, and dismissing the notion of mystic powers at work behind the scenes. But just one of his great insights is relevant to my present story. What Descartes had realized, while lying in bed and watching the erratic movement of a fly buzzing around in the corner of his room, was that the position of the fly at any moment in time could be defined simply by three numbers, specifying the distance of the fly from each of the three surfaces (two walls and a ceiling) that met in the corner. Although he immediately saw this in three-dimensional terms, we are more used to thinking of such coordinate systems, as they are called, in two dimensions, on the surface of the Earth or on a piece of graph paper. The notion that a point on a piece of paper can be specified using two numbers, x and y, is, indeed, so ingrained in us that the surprise today is that anybody had to come up with the idea at all. But Descartes did, and in his honour such systems of measurement of position are known as Cartesian coordinates. If you ever give directions to someone in a modern city by telling them to go 'three blocks north and two blocks east' to find their destination, or if you locate the position of their objective on a town map in terms of numbered squares, you are giving the directions in Cartesian coordinates (Figure 2.1).

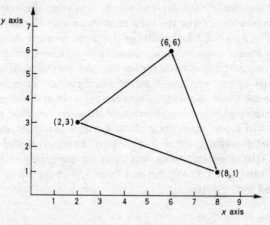

Figure 2.1 *Using Cartesian coordinates, three pairs of numbers – (2,3), (6,6) and (8,1) – completely specify a particular triangle.*

In the same way, you can specify the properties of a geometrical shape in terms of Cartesian coordinates. Once you choose a set of axes as a reference (usually two lines at right angles to each other, like the x and y axes of a graph), a triangle, for example, can be represented by three pairs of numbers that specify the positions of its vertices. So Descartes opened up the possibility of using relationships between sets of numbers – algebraic equations – to study geometry. He also opened up the possibility of turning problems in algebra into problems in geometry, by representing equations as graphs.* And all this is by no means restricted to shapes made up of straight lines, like triangles. Any curved line, in two dimensions, can be defined by setting out the pairs of numbers (the xs and ys) that specify each point along the line, or by an equation which tells you how to work out the xs and ys. The same

---

* My father used to tell me that he had only ever passed his maths exams in school (back in the 1930s) by literally converting algebraic problems into graphs, and measuring off the answers he required, instead of solving the equations. If so, it was thanks to Descartes that he passed those exams!

thing applies in three dimensions (for example, to calculate the flight path of a fly) provided that you have three reference axes and three coordinate numbers.

Where do you measure the coordinate numbers from? It doesn't matter! You can put the base you measure from (the origin of the axes, in the case of those graphs we drew in school) anywhere you like; you can even twist the axes around, to alter the direction (or directions) they point in, or change the angle between them, so that it is no longer a right angle. There will still be a unique set of Cartesian coordinates to describe the line you are interested in. And, just as you can use a set of numbers (or an equation) to describe the shape of a line, so you can use a set of numbers (or the appropriate equations) to describe the shape of a surface, such as a flat sheet of paper, the surface of the Earth, a soft drink can, or (in principle) something more complicated, like a crumpled piece of paper. That is exactly what was done by the nineteenth-century mathematicians who went beyond Euclid, with the aid of the tools provided by Descartes.

Beyond Euclid

The first person to go beyond Euclid and to appreciate the significance of what he was doing was the German Karl Gauss, one of the greatest of all mathematicians, who was born in Brunswick in 1777. He came from a poor family (his father was a gardener and an assistant to a local merchant), but displayed such a remarkable talent for mathematics that at the age of fourteen he was presented at court, by friends of his schoolteacher, to the Duke of Brunswick. Brunswick became his patron and supported Gauss financially, until the Duke died of wounds received while fighting against the army of Napoleon at the battle of Jena in 1806. By then, Gauss was not only well established in his own right, but had already, at the age of twenty-nine, completed almost all of his great contributions to mathematics. Much of his work, though, was still unknown to other scientists, let alone the world at large.

There were two reasons for this. First, Gauss made many important discoveries in mathematics between the ages of fourteen and seventeen, when, under the Duke's patronage, he was studying at the Collegium Carolinium in Brunswick. The young genius from a poor family simply did not know how to set about having his work published. From 1795 to 1798, Gauss studied at the University of Göttingen, and continued to make new discoveries in mathematics; by the time he received his doctorate, from the University of Helmstedt in 1799, at the age of twenty-two, his greatest mathematical accomplishments had all been achieved. The second reason why Gauss published relatively little of his work, even after he had become familiar with the workings of the academic world, is that he was a perfectionist. He would only publish something if he had developed a complete account of the discovery and its implications, polished and honed to his own satisfaction. As a result, many important discoveries in mathematics made by other researchers in the nineteenth century later turned out to have been pre-empted by Gauss, who had left them, unpublished, in his notebooks.

By the beginning of the nineteenth century, Gauss had turned the main focus of his scientific attention to astronomy, and after the death of the Duke of Brunswick he became the Director of the Göttingen Observatory, as well as being a professor at the university, where he remained until he died in February 1855. His notebooks, written in his own form of mathematical shorthand, contain some entries that have never been deciphered, and may refer to mathematical discoveries that later generations have as yet still failed to reproduce; but they also show that he discovered one form of non-Euclidean geometry in 1799, exactly thirty years before the first description of such a geometry was actually published, by the Russian mathematician Nikolai Ivanovitch Lobachevsky.

Lobachevsky (who first discussed the idea publicly in 1826) had also been pre-empted by a Hungarian army officer, János Bolyai. Bolyai was not a complete amateur, but the son of another mathematician, Wolfgang, who was a contemporary and friend of Gauss. Wolfgang had wanted János to study under Gauss in Göttingen, but had been disappointed to see the young man join the army in 1818, at the age of sixteen. Like Descartes, the younger Bolyai was not a fighting soldier but an engineering officer; encouraged to probe the nature of Euclidean geometry by his father's obsession with the parallel postulate, he had made a similar discovery to that of Lobachevsky and Gauss in 1823, but failed to get it published until 1832.

All three researchers hit on essentially the same kind of 'new' geometry. They showed that it is possible to set up a complete, self-consistent geometry in which all the axioms and postulates of Euclidean geometry hold, *except* the parallel postulate. In the specific non-Euclidean geometry developed by each of them independently, it is possible to draw a straight line and mark a point not on that line through which you can draw many more lines, none of which crosses the first line, and which are therefore all parallel to it. This kind of geometry applies on a surface curved in a particular way, and known as a 'hyperbolic' surface. It is shaped like a saddle – such a surface is open, and extends to infinity (Figure 2.2). On an

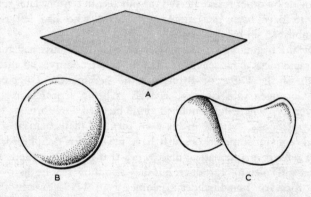

Figure 2.2 *The Euclidean geometry we learned at school only works perfectly on flat surfaces (A). You need different rules of geometry to describe what goes on on curved surfaces, which may be either closed (B) or open (C). Three-dimensional space can also be curved; our Universe is very nearly flat, but is almost certainly slightly curved, like the surface of a sphere, so that it is closed. A black hole closes space off around itself in this way.*

open, hyperbolic surface, the sum of the angles of a triangle is always *less* than 180°; it is described as having negative curvature.

Most people find it easier to grasp the concept of non-Euclidean geometry from a different example, which, curiously, was not the way the three non-Euclidean pioneers made their discovery. This is the surface of a sphere, like the surface of the Earth, which does not extend to infinity (always an uncomfortable concept for non-specialists) and is said to be closed, or to have positive curvature. It is easy to see that parallel lines behave strangely on the surface of a sphere – take any two lines of longitude, that start out straight and parallel at the equator, running due north and south; like all other lines of longitude, they will cross each other twice, at the north and south poles. On a closed surface like this, the angles of a triangle always add up to more than 180° (Figure 2.3). Flat surfaces, where

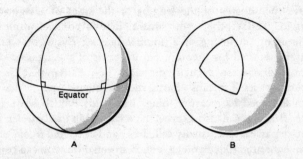

A   **B**

Figure 2.3 *A. On a sphere, lines of longitude all cross the equator at right angles, so they are parallel – but they all meet at the poles!*
B. And the angles of a triangle drawn on the surface of a sphere add up to more than 180°.

Euclidean geometry holds sway, are, after all, just a special case on the borderline between the open and shut possibilities. But the fact that there are *many* possible non-Euclidean geometries, both open and closed, did not sink in to the minds of the mathematicians until the work of Bernhard Riemann, a pupil of Gauss, in the 1850s. It was Riemann who, among other things, discovered 'spherical' geometry.

Geometry comes of age

Riemann opened up the worlds of non-Euclidean geometry by treating them algebraically, in Cartesian terms. This offers a literal infinity of possibilities, which cannot be dealt with by geometers restricted to using measuring sticks, protractors, compasses and so on. Those instruments work well enough if you are investigating the relationships between different shapes drawn on a two-dimensional surface, or even between different objects in three-dimensional space. But how can you measure things in four (or more) dimensions? You can't, and only a mathematician would even think of asking the question. But you *can* write down, and manipulate, the equations which describe such multi-dimensional phenomena.

Take the famous theorem of Pythagoras, for example, which relates the squares of the lengths of the sides of a right-angled triangle. Today, the term 'square' in this context immediately conjures up an image of a number like x^2; Pythagoras himself, though, worked his theorem out in terms of the areas of squares literally drawn on each of the sides of the triangle. The term 'geometry' itself means 'Earth measurement', and it developed from the need to measure things like fields, which have definite areas. But you can also express the relationship in Cartesian terms, as three parameters (usually called x, y and z) related to one another by an equation. Once you have such an equation, you can construct similar equations with more than three parameters (indeed, as many parameters as you like) which are related to one another by the same rule that gives us Pythagoras' theorem for a triangle. In some sense, the equation with the extra terms is describing the geometry of the equivalent of triangles in higher-dimensional space.

All of this is fascinating for mathematicians, if not of much interest to lesser mortals (except, as we shall soon see, when it comes to describing the geometry of four-dimensional space). And it was Bernhard Riemann who pointed out the mathematical possibilities.

Riemann, who had been born in 1826, entered Göttingen University at the age of twenty, and learned his mathematics initially from Gauss, who had turned seventy by the time Riemann moved on to Berlin in 1847, where he studied for two years before returning to Göttingen. He was awarded his doctorate in 1851, and worked for a time as an assistant to the physicist Wilhelm Weber, an electrical pioneer whose studies helped to establish the link between light and electrical phenomena, partially setting the scene for Maxwell's theory of electromagnetism. The accepted way for a young academic like Riemann to make his way in a German university in those days was to seek an appointment as a kind of lecturer known as a *Privatdozent*, whose income would come from the fees paid by students who voluntarily chose to take his course (an idea which it might be interesting to revive today). In order to demonstrate his suitability for such an appointment, the applicant had to present a lecture to the faculty of the university, and the rules required the applicant to offer three possible topics for the lecture, from which the professors would choose the one they would like to hear. It was also a tradition, though, that although three topics had to be offered, the professors always chose one of the first two on the list. The story is that when Riemann presented his list for approval, it was headed by two topics which he had already thoroughly prepared, while the third, almost an afterthought, concerned the concepts that underpin geometry. Riemann was certainly interested in geometry, but apparently he had not prepared anything along these lines at all, never expecting the topic to be chosen. But Gauss, still a dominating force in the University of Göttingen even in his seventies, found the third item on Riemann's list irresistible, whatever convention might dictate, and the twenty-seven-year-old would-be *Privatdozent* learned to his surprise that that was what he would have to lecture on to win his spurs.

Perhaps partly under the strain of having to give a talk he had not prepared and on which his career depended, Riemann fell ill, missed the date set for the talk, and did not recover until after Easter in 1854. He then prepared the lecture over a period of seven weeks, only for Gauss to call a postponement on the grounds of ill health. At last, the talk was delivered, on 10 June 1854. The title, which had so intrigued Gauss, was 'On the hypotheses which lie at the foundations of geometry'. In that lecture – which was not published until 1867, a year after Riemann died – he covered an enormous variety of topics, including a workable definition of what is meant by the curvature of space and how it could be measured, the first description of spherical geometry (and even the speculation that the space in which we live might be gently curved, so that the entire Universe is closed up, like the surface of a sphere, but in three dimensions, not two), and, most important of all, the extension of geometry into many dimensions with the aid of algebra.

Although Riemann's extension of geometry into many dimensions was the most important feature of his lecture, the most astonishing, with hindsight, was his suggestion that space might be curved into a closed ball. This is even more remarkable than the idea of dark stars arrived at by Michell and by Laplace, since, after all, they were simply applying Newtonian ideas about gravity to the Newtonian concept of light as tiny particles. More than half a century before Einstein came up with the general theory of relativity – indeed, a quarter of a century before Einstein was even born – Riemann was describing the possibility that the entire Universe might be contained within what we would now call a black hole. 'Everybody knows' that Einstein was the first person to describe the curvature of space in this way – and 'everybody' is wrong.

Of course, Riemann got the job – though not because of his prescient ideas concerning the possible 'closure' of the Universe. Gauss died in 1855, just short of his seventy-eighth birthday, and less than a year after Riemann gave his classic exposition of the hypotheses on which geometry is based. In 1859, on the death of Gauss's successor, Riemann himself took over as professor, just four years after the nerve-wracking experience of giving the lecture upon which his job as a humble *Privatdozent* had depended (history does not record whether he ever succumbed to the temptation of asking later applicants for such posts to lecture on the third topic from their list). He died, of tuberculosis, at the age of thirty-nine. If he had lived as long as Gauss, he would have seen his intriguing mathematical ideas about multi-dimensional space begin to find practical applications in Einstein's new description of the way things move. But Einstein was not even the second person to think about the possibility of space in our Universe being curved, and he had to be set out along the path that was to lead to the general theory of relativity by mathematicians more familiar with the new geometry than he was.

The geometry of relativity

Chronologically, the gap between Riemann's work and the birth of Einstein is nicely filled by the life and work of the English mathematician William Clifford, who lived from 1845 to 1879 and, like Riemann, died of tuberculosis. Clifford translated Riemann's work into English, and played a major part in introducing the idea of curved space and the details of non-Euclidean geometry to the English-speaking world. He knew about the possibility that the three-dimensional Universe we live in might be closed and finite, in

the same way that the two-dimensional surface of a sphere is closed
and finite, but in a geometry involving at least four dimensions.
This would mean, for example, that just as a traveller on Earth who
sets off in any direction and keeps going in a straight line will
eventually get back to their starting point, so a traveller in a closed
universe could set off in any direction through space, keep moving
straight ahead, and eventually end up back at their starting point.
But Clifford realized that there might be more to space curvature
than this gradual bending encompassing the whole Universe. In
1870 he presented a paper to the Cambridge Philosophical Society
(at the time, he was a Fellow of Newton's old college, Trinity) in
which he described the possibility of 'variation in the curvature of
space' from place to place, and suggested that 'small portions of
space *are* in fact of nature analogous to little hills on the surface [of
the Earth] which is on the average flat; namely, that the ordinary
laws of geometry are not valid in them.' In other words, still seven
years before Einstein was born, Clifford was contemplating *local*
distortions in the structure of space – although he had not got
around to suggesting how such distortions might arise, nor what
the observable consequences of their existence might be.

Clifford was just one of many researchers who studied non-
Euclidean geometry in the second half of the nineteenth century* –
albeit one of the best, with some of the clearest insights into what
this might mean for the real Universe. His insights were particularly profound, and it is tempting to speculate how far he might
have gone in pre-empting Einstein if he had not died eleven days
before Einstein was born. Curiously, while Will Clifford can
almost be regarded, with hindsight, as a pioneering relativist of the
1870s, one of the highly respected relativists of the 1970s and 1980s,
and the author of the best introduction for the lay person to general
relativity,† is the American researcher Clifford Will, who was born
101 years after his inverse namesake.

With all of this interest in geometry in the latter part of the
nineteenth century, it may seem odd that Einstein arrived at his
special theory of relativity, and presented it to the world in 1905,
using purely algebraic techniques, setting out the equations that
describe motion in a way consistent with Maxwell's discovery that

---

* A comprehensive guide to their activity has been given by J. D. North, in *The Measure of
the Universe*, Oxford University Press, 1965.
† *Was Einstein Right?*, Basic Books, 1986.

the speed of light is a constant, and solving them. But, of course, Einstein was a physicist, not a mathematician. He was very nearly not even a physicist; he was so bored by the dull methods of his teachers that he was told by several of them that he would never amount to much, was expelled from one school in Germany, failed the entrance examination to the polytechnic in Zürich the first time he took it, and even after cramming for a year and passing the examination at his second attempt was later described by one of his teachers there, Hermann Minkowski, as a 'lazy dog' who was certainly quite bright, but who 'never bothered about maths at all'. It wasn't only maths that he didn't bother with. When the time approached for the final examinations, Einstein was way behind in many subjects, having skipped all the lectures that he found boring. He was again forced to do some hard cramming in order to catch up, and passed only with the aid of the lecture notes taken down by his friend Marcel Grossman, a more assiduous student who went on to a distinguished career in science in his own right. The story has often been told of how Einstein scraped his degree (graduating in 1900) but failed to get an academic job, and spent the early years of the twentieth century working in the patent office in Bern – a job arranged for him through the good offices of Grossman's father. The job was sufficiently undemanding to leave Einstein plenty of free time to think about physics. The special theory of relativity was published in 1905, and the rest is history.

Well, almost. Special relativity was not an immediate sensation that swept the board and established Einstein's reputation. Indeed, partly from choice – he was offered other jobs – he stayed in the patent office until 1909, when his steadily growing reputation brought him the offer of a job at the University of Zürich. Part of the reason why his reputation had grown so much by then was that his old teacher, Hermann Minkowski, had taken Einstein's algebraic expression of the special theory and developed a geometrical description in four dimensions which improved the clarity of the theory and is still regarded as the best way to understand it. It was Minkowski who put the geometry into relativity.

Minkowski had been born back in 1864, two years before Riemann died. He held the post of Professor of Mathematics at the polytechnic in Zürich (its full name is the Federal Institute of Technology) for only half a dozen years (overlapping with Einstein's time as a student there); from 1902 until his death from appendicitis in January 1909, he worked at Göttingen University, following in the tradition of Gauss and Riemann. But his geometrization of the special theory of relativity owed as much to Descartes as to either of his eminent predecessors in Göttingen.

Einstein's equations of motion – the equations of the special theory of relativity – involve four parameters, which describe the location of an object in terms of the usual three coordinates of three-dimensional space, plus another coordinate representing time. Remember Descartes, lying in his bed and watching the fly buzzing around the corner of his room. He realized that its position *at any moment in time* could be specified by three spatial coordinates. In effect, Einstein's equations said that the entire life history of the fly could be specified in terms of *four* Cartesian coordinates, three of space and one of time. You can imagine drawing a line tracing the route followed by the fly through space from the moment the egg it emerged from was laid to the moment it died, a very wiggly line that, on one particular day, 10 November 1619, happened to pass through Descartes' bedroom. Such a line is now called a world line, and it exists in four dimensions (Figure 2.4).

One of the key equations of Einstein's special theory of relativity actually looks rather like the equation that describes the algebraic version of Pythagoras' theorem. This is no coincidence. The equation tells how to identify, or measure, the shortest distance between two points. Such an equation is said to describe the metric of the multidimensional space (or spacetime), from the same root as the 'metry' in 'geometry'; the shortest distance between two points in any multidimensional space is itself called a geodesic, and, of course, on a flat sheet of paper, or the mud of the fields near the River Nile, the geodesics are simple straight lines and the metric is described by Pythagoras' theorem. It works like this. In two dimensions, if we identify two points by Cartesian coordinates x and y, we can then draw a right-angled triangle and calculate the shortest distance between those points (which is the hypotenuse of the triangle) using Pythagoras (Figure 2.5). We can do the same thing in three dimensions, using three coordinates, x, y and z. And, Minkowski realized, Einstein's equations said that we can do the same thing in *four* dimensions, using four coordinates, x, y, z and t – in everyday language, by specifying the location in terms of 'up/down', 'left/right', 'forward/backward' *and* 'past/future'. Minkowski's geometrization of special relativity is a combination of

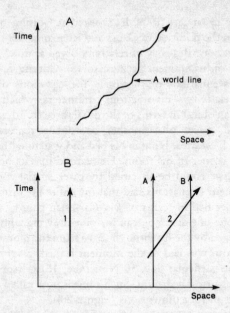

Figure 2.4 *A. A 'spacetime diagram' portrays how things move. The three dimensions of space are represented by the 'x-axis', and the passage of time by the 'y-axis'. The world line of an object (a fly, perhaps) shows its position in space at any instant of time.*
B. Particle 1 stays in the same place all the time. Its world line is vertical. Particle 2 moves from A to B as time passes. It has a sloping world line.

Descartes' insight and Riemann's extension of geometry into four dimensions.*

This provides the clearest explanation of how time can run slower and rulers can shrink when you move at speeds close to the speed of light. Einstein's equations tell us that there is an equivalent of length, measured in four dimensions; this is called extension.

* When we are dealing with curved surfaces (or spaces), we can still construct tiny Pythagorean triangles and measure the sizes of the squares on each of their sides, but now these squares (or areas) will not obey the rule Pythagoras discovered. The way in which the squares deviate from Pythagoras' rule then provides us with a measure, in metric terms, of the nature (open or closed) and size of the curvature of space (or even of spacetime). But this, of course, is exactly what Einstein was still *not* aware of in 1909.

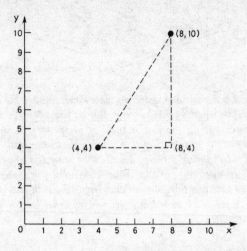

Figure 2.5 *If we know the Cartesian coordinates of two points, we can make a right-angled triangle and use Pythagoras' theorem to calculate the shortest distance between the two points. This works wherever we put the zero of our coordinate system – wherever we measure x and y from – because what matters is the differences between the pairs of numbers. The same trick works in four (or more!) dimensions; even though we cannot draw four-dimensional 'triangles', the equations corresponding to Pythagoras' theorem in four dimensions are easy to write down and work out. So we can calculate the shortest distance between points in spacetime, not just in space.*

The extension of, say, a ruler can be thought of as the length of the hypotenuse of a four-dimensional triangle, calculated using Pythagoras' theorem, and it does not change. But to a moving observer, the perspective of this extension *does* change, with time stretching and length shrinking, in a perfect balance with one another.

Think of a drum majorette's mace, in the familiar three dimensions of everyday space. It always has the same length. But depending on your point of view – your perspective – it may seem to be very short, because you are viewing it nearly end on, or it may show its full length, if you are viewing it from the side. As it twirls through the air, its length seems to be constantly changing – but it is all a matter of perspective (Figure 2.6). And that is how you

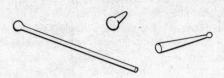

Figure 2.6 As a drum majorette's mace twirls through the air, it looks as if its length keeps changing. But we know from everyday experience that this is just a trick of perspective. Einstein's discovery of the strange way that moving clocks run slow while moving rulers shrink actually describes the same sort of perspective effect operating in the four *dimensions of spacetime. Although time and three-dimensional length are both distorted by motion, there is an underlying 'four-dimensional length' which stays the same.*

explain the strange features of special relativity using geometry – in terms of varying perspective in *four* dimensions, three of space and one of time.

There is one major, and one minor, additional subtlety. In the equations, the parameter involving time appears with a minus sign, while the three parameters representing the three dimensions of space each appear with a plus sign. This is why time cannot be regarded simply as a fourth dimension of space. It *is* a fourth dimension, but it is a kind of *negative* space. When rulers shrink, time expands; when rulers expand, time shrinks. But all the while, the four-dimensional extension of the ruler in spacetime stays the same. In addition, the parameter representing time in the equations is always multiplied by the speed of light, so that one second of time is equivalent to just under 300,000 kilometres of space. This is why relativistic effects become noticeable only if you are moving at a sizeable fraction of the speed of light.

Minkowski's enormous simplification of special relativity was described in a lecture he gave in Cologne in 1908, and appeared in print in 1909, shortly after his death. His opening words in that lecture indicate the importance he attached to this new concept of four-dimensional spacetime, an importance that was immediately recognized by others:

The views of space and time which I wish to lay before you have sprung from the soil of experimental physics, and therein lies their strength. They are radical. Henceforth space by itself, and time by itself, are doomed to fade into mere shadows, and only a kind of union of the two will preserve an independent reality.*

But one of the people who was not, at first, impressed by Minkowski's geometrization of special relativity was that well-known 'lazy dog', Albert Einstein, who never had taken much interest in mathematics. He soon learned to live with it, though, no doubt encouraged by the fact that his reputation began to take off almost immediately from the point in spacetime where Minkowski uttered those words – as shown, for example, by the fact that Einstein was awarded an honorary doctorate, the first of many, by the University of Geneva in July 1909.

Even so, although Minkowski's version of special relativity took on board one of Riemann's many ideas, the notion of multi-dimensional geometry, it took no account at all, because it had no need to, of Riemann's even more profound ideas concerning curved space. The geometry of the special theory of relativity is still Euclidean geometry, obeying the rules which apply in flat space. It is just that it is Euclidean geometry extended to four dimensions – to flat *spacetime*. The next big step came when Einstein, prodded hard by his old friend Grossman, began to consider the implications of *curved* spacetime, going beyond the special case of flat spacetime to create a more general theory – the general theory of relativity.

Einstein's gravitational insight

Conventional wisdom has it that although special relativity was a product of its time, and had Einstein not come up with the theory in 1905 someone else soon would have, under the pressure of the need to explain the conflict between Newtonian mechanics and the behaviour of light, general relativity was a work of unique inspiration, which sprang from Einstein's genius alone, and which might not have been discovered for another fifty years if he had fallen under the wheels of a tram in 1906. I have even been guilty of perpetuating this myth myself, in earlier books. But now it seems to me that this case does not stand up to close inspection. It is a case made by physicists, looking back at how Einstein's theory describes material objects. The conflict between Newton and Maxwell

---

* Quoted by Abraham Pais, in *Subtle is the Lord*.

pointed to a need for a new theory, but once that theory was in place, the argument runs, there were no outstanding observational conflicts that had still to be explained. Maybe. But by the 1900s, as I have pointed out, many mathematicians were already intrigued by the notion of curved space. Once Minkowski had presented special relativity as a theory of mechanics in *flat* four-dimensional spacetime, it would, surely, not have been long before somebody (maybe even Grossman) wondered how those laws of mechanics would be altered if the spacetime were curved. From a mathematical point of view, general relativity is every bit as much a child of its time as special relativity was, and a logical development from the special theory (indeed, as we shall see, in this connection the mathematicians stayed several jumps ahead of the physicists until well into the 1960s, and are still a jump or two ahead even today). This is certainly borne out by the fact that it took the prodding of a mathematician to get the physicist Einstein moving along the right lines after 1909.

What Einstein lacked in terms of top-flight mathematical skill and knowledge, though, he more than made up for in terms of physical intuition – his 'feel' for the way the Universe worked was second to none. His special theory of relativity, for example, developed from Einstein wondering what the Universe would look like if you could ride with a light ray as it hurtled through space at nearly 300,000 km a second; and the seed that grew into the general theory was an inspired piece of reasoning about the behaviour of a light ray crossing a falling elevator. The seed was sown within a couple of years of the completion of the special theory; but, partly because at that time Einstein knew nothing of Riemannian geometry, it took a further nine years to grow to fruition.

The special theory of relativity tells us how the world looks to observers moving with different velocities. But it deals only with constant velocities – steady motion at the same speed in the same direction. Even in 1905, it was obvious that the theory failed to describe how objects behave under two important sets of conditions that exist in the real world. It does not describe the behaviour of accelerated objects (by which physicists mean, remember, objects that change their speed *or* their direction, or both); and it does not describe the behaviour of objects that are under the influence of gravity. Einstein's insight, which he first presented in 1907, was that both these sets of conditions are the same – that acceleration is exactly equivalent to gravity. This is such a cornerstone of our modern understanding of the Universe that it is known as the 'principle of equivalence'.

Anyone who has travelled in a high-speed elevator knows what Einstein meant by the principle. When the elevator starts moving upward, you are pressed to the floor, as if your weight has increased; when it slows at the top of its rise, you feel lighter, as if gravity has been partly cancelled out. Clearly, acceleration and gravity have something in common; but it is a dramatic step to go from this observation to say that gravity and acceleration are *exactly the same*. An implausible scenario demonstrates just how equivalent they are. If the cable of the elevator snapped and all the safety devices failed, while the lift was falling freely down its shaft you would fall at the same rate, weightless, floating about inside the falling 'room'.

But what would happen to a beam of light shone across the falling elevator from one side to the other? In the weightless falling room, according to Einstein, Newton's laws apply and the light must travel in a straight line from one side to the other. Then, however, he went on to consider how such a beam of light would look to anyone outside the falling elevator, if the lift had walls made of glass and the path of the light beam could be tracked. In fact, the 'weightless' elevator and everything inside it is being accelerated by the gravitational pull of the Earth. In the time it takes the light beam to cross the elevator, the falling room has increased its speed, and yet the light beam still strikes the spot on the opposite wall that is level (according to an observer in the lift) with the spot from where it started. This can only happen if, from the point of view of the outside observer, the light beam has bent downward slightly while crossing the falling elevator. And the only thing that could be doing the bending is gravity.

So, said Einstein, if acceleration and gravity are indeed *precisely* equivalent to one another, gravity must bend light. You can cancel out gravity while you are in free fall, constantly accelerating; and you can create an effect indistinguishable from gravity by providing an acceleration, which makes everything 'fall' to the back of the accelerating vehicle (Figure 2.7).

The possibility of light bending was neither new nor startling – as we have seen, Newtonian mechanics and the corpuscular theory suggest that light should be bent, for example when it passes near the Sun. Indeed, Einstein's first calculations of gravitational light bending, based on the principle of equivalence, suggested that the

Figure 2.7 *Gravity and uniform acceleration produce identical forces, which we call weight.*

amount of bending would be exactly the same as in the old Newtonian theory. Fortunately, though, before anyone could carry out a test to measure the predicted effect (not that anyone was very interested in it while the theory was incomplete), Einstein had developed a full theory of gravity and accelerations, the general theory of relativity. In the general theory, the predicted light bending is *twice* as much as in the Newtonian version, and it was the measurement of this non-Newtonian effect that made people sit up and take notice of the general theory. But that wasn't until 1919.

For more than three years after he first stated the principle of equivalence, Einstein did very little work on trying to develop a proper theory of gravity based on the principle. There were many reasons for this. As Einstein's reputation grew, he took up a series of increasingly prestigious academic posts, first as a *Privatdozent* in Bern, then as assistant professor in Zürich, then on to be a full professor in Prague. He had a growing family – his son Hans had been born in 1904, and Eduard arrived in 1910. But, most important of all, during that period Einstein's scientific attention was focused on his contributions to the exciting new developments in quantum physics, and he simply didn't have time to struggle with a new theory of gravity as well. It was after he had reached a temporary impasse with his work on quantum theory that, in Prague in the summer of 1911, he returned to the gravitational fray.

The relativity of geometry

It was in 1911, in fact, that Einstein first applied the idea of light bending to rays passing close by the Sun, and came up with a prediction essentially the same size as the Newtonian prediction. The Newtonian version of the calculation had been made back in 1801, by the German Johann von Soldner, acting on the assumption that light is a stream of particles; Einstein, completely unaware of von Soldner's calculation, calculated his own initial version of light bending by the Sun in 1911 by treating light as a wave (even though he had himself been instrumental in showing that light sometimes *does* behave like a stream of particles!). The two calculations give almost precisely the same value for the bending. The simplest way to understand the first Einsteinian version of the effect is that it results from the distortion of time caused by the Sun's gravitational field. In 1911, Einstein was struggling with a horribly complex and unwieldy set of equations that in effect corresponded to a combination of warped time with flat space, and as a result he was literally only halfway to the full value of the light bending effect.

Things began to look up, however, as soon as Einstein moved back to Zürich, after staying in Prague for only a year. His return to Switzerland was engineered by the friend whose lecture notes he had borrowed a dozen years before – Marcel Grossman, who had now risen to become dean of the physics and mathematics department of the polytechnic.

Grossman's own career had followed a much more conventional pattern than Einstein's, although he had reached this eminence very young. He was just one year older than Einstein, and after graduating with Einstein in 1900 he worked as a teacher while writing his doctoral thesis, also producing two geometry books for high-school students and several papers on non-Euclidean geometry. On the strength of this work, he joined the faculty at the polytechnic, becoming a full professor in 1907 and dean in 1911 at the age of thirty-three. One of his first acts as dean was to entice

Einstein back to Zürich. He arrived on 10 August 1912, knowing that he had the basis of a workable theory of gravity, but uncomfortably aware that he lacked the right mathematical tools to finish the job. Much later, he recalled a plea he made at this time to his old friend – 'Grossman, you must help me or I'll go crazy!'[*] Einstein had realized that the method for describing curved surfaces developed by Gauss (essentially, the metric technique I have mentioned) might help with his difficulties, but he knew nothing about Riemannian geometry. He did, however, know that Grossman was a whizz at non-Euclidean geometry, which is why he turned to him for help – 'I asked my friend whether my problem could be solved by Riemann's theory.' The answer, in a word, was yes. Although it took a long time to sort out the details, what Grossman was able to tell Einstein immediately opened the door for him, and by 16 August he was able to write to another colleague 'it is going splendidly with gravitation. If it is not all deception, then I have found the most general equations.'

Einstein and Grossman investigated the significance of curved spacetime (warping *both* space *and* time) for a theory of gravity in a paper published in 1913. The collaboration ended when Einstein accepted an appointment as director of the new Institute of Physics at the Kaiser Wilhelm Institute in Berlin in 1914 – a post so tempting, requiring no teaching and allowing him to devote all his time to research, that it tore him away from Switzerland and Grossman. But the two remained firm friends until Grossman's death, from multiple sclerosis, in 1936. It was in Berlin that Einstein, alone, completed the long journey from the special theory of relativity to the general theory.

The full version of the general theory was presented at three consecutive meetings of the Prussian Academy of Sciences in Berlin in November 1915, and published in 1916. Although it has implications far beyond the scope of the present book, what matters here is the way in which Einstein used Riemannian geometry to describe curved space. A massive object, like the Sun, can be thought of as making a dent in three-dimensional space, in a way analogous to the way an object like a bowling ball would make a dent in the two-dimensional surface of a stretched rubber sheet, or a trampoline. The shortest distance between two points on such a curved surface will be a curved geodesic, not what we are used to

---

[*] Quotations in this section are from Abraham Pais, *Subtle is the Lord*.

thinking of as a straight line, and this applies in the three-dimensional case as well. Because space is bent, light rays are bent (Figure 2.8). But Einstein had already discovered, as we have seen,

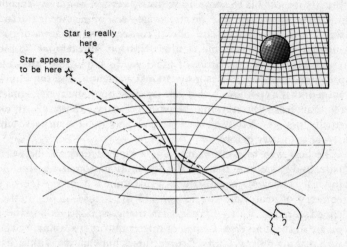

Figure 2.8 *A heavy object placed on a stretched rubber sheet (*inset*) makes an indentation. The presence of the Sun 'indents' spacetime in an analogous way. This accounts for the light-bending effect described on p. 29 (Figure 1.3).*

that light rays are bent near a massive object by a warp in the time part of spacetime, as well. And, as it happens, the space warping alone bends the light by the same amount as the time-warping effect that Einstein had already calculated. Overall, the general theory of relativity predicts *twice* as much light bending as Newtonian theory does.★ That is why, when the light bending was measured during the eclipse of 1919 and found to agree with Einstein, not Newton, the newspapers proclaimed that Newton's theory of gravity had been overthrown. But that is wrong.

What Einstein had actually done was to *explain* Newton's law of gravity. There are some subtle differences, such as with the bending

★ Indeed, it is the 'new' space-warping effect discussed by Einstein in 1916 that is actually the equivalent of the old Newtonian effect; it is the *time* warping that makes the relativistic prediction different from the Newtonian calculation.

of light by the Sun, between simple Newtonian theory and the general theory of relativity. But what really matters is that if gravity is explained as the result of curvature in four-dimensional spacetime, then, because of the nature of this curvature itself, it is virtually impossible to come up with any version of gravity except an inverse square law. An inverse square law of gravity is far and away the most natural, and likely, consequence of curvature in four-dimensional spacetime. Unlike Newton, Einstein *did* 'frame hypotheses' about the nature of gravity. His hypothesis was that spacetime curvature causes gravitational attraction, and the implication of that hypothesis is that gravity must obey an inverse square law. Far from overturning Newton's theory, Einstein's work actually *explains* Newton's theory, and puts it on a more secure footing than ever before.

The best way to picture this is as a kind of dialogue between matter and spacetime. Because the distribution of matter across the Universe is uneven, the curvature of spacetime is uneven – the very geometry of spacetime is relative, and the nature of the metric, defined in terms of tiny Pythagorean triangles, depends on where you are in the Universe. Lumps of matter distort spacetime, not so much making hills, as Clifford conjectured, but valleys. Within that curved spacetime, moving objects travel along geodesics, which can be thought of as lines of least resistance. And you can calculate the length of even a curved geodesic in general relativity in terms of many tiny Pythagorean triangles which each 'measure' a tiny portion of its length, added together using the integral calculus developed originally by Newton. But a falling rock, or a planet in its orbit, doesn't have to make the calculation – it just does what comes naturally. In other words, matter tells spacetime how to bend, and spacetime tells matter how to move.

There is, however, one important point about all this which often causes misunderstandings and confusion. We are *not* just dealing with curved *space*. The orbit of the Earth around the Sun, for example, forms a closed loop in space. If you imagined that this represents the curvature of space caused by gravity, you would leap to the false conclusion that space itself is closed around the Sun – which it obviously is not, since light (not to mention the Voyager space probes) can escape from the Solar System. What you have to remember is that the Earth and the Sun are each following their own world lines through four-dimensional spacetime. Because the factor of the speed of light comes into the time part of Minkowski's

metric for spacetime, and this carries over into the equivalent
metric in general relativity, these world lines are enormously
elongated in the time direction. So the actual path of the Earth
'around' the Sun is not a closed loop, but a very shallow helix, like
an enormously stretched spring (Figure 2.9). It takes light eight and

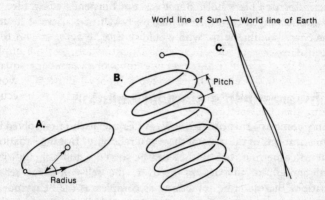

World line of Sun — ← World line of Earth

Pitch

Radius

Figure 2.9 A. *Earth's orbit around the Sun is closed in ordinary
space.*
B. In spacetime, the Earth's orbit is like a coiled spring, or helix.
*C. Actually, because the speed-of-light factor is so big, the helix
is so elongated that the Earth's world line is very nearly straight –
the 'pitch' of the helix is 63,000 times its radius!*

one third minutes to reach the Earth from the Sun. So each circuit
that the Earth makes around the Sun is a distance of about 52 light
minutes. But it takes a year for the Earth to complete such a circuit,
and in that time it has moved along the time direction of spacetime
by the equivalent of a light year – more than ten thousand times
further than the length of its annual journey through space, and
more than 63,000 times the distance from the Earth to the Sun. In
other words, the pitch of the helix representing the Earth's journey
through spacetime is more than 63,000 times bigger than its radius.
In flat spacetime, the world line would be a straight line; the
presence of the Sun's mass actually distorts spacetime only slightly,
just enough to cause a slight bending of the world line, so that it
weaves to and fro, very gently, as the Earth moves through

spacetime. You need to have much more mass, or a much higher density of mass, in order to close space around an object.

Within a few weeks of his presentation of the general theory, Einstein was back at the Prussian Academy of Sciences, reporting a solution to his equations that described just such a phenomenon. This was the first public airing of the correct mathematical description of a black hole. But it was not Einstein's work. He was communicating it to the Academy on behalf of a scientist he had been corresponding with, who would shortly lie dying in a hospital in Potsdam.

Schwarzschild's singular solution

It may come as a surprise that it wasn't Einstein who first solved his own equations of the general theory of relativity. But the equations have to come first, before they can be solved – and although it is hard enough to invent, or discover, the self-consistent set of equations that describe something as complex as the behaviour of spacetime in the presence of matter, even that is no guarantee that the equations will be easy to solve. In a way, it's a bit like a crossword puzzle. The person who compiles a crossword puzzle knows what words are supposed to fit into the squares, and has some freedom to choose the grid of squares that they want to make the words fit; but even then it is no easy task to make the puzzle work, if you obey the strict rules that puzzle compilers follow, in particular the requirement that the pattern of squares should be symmetrical. Once the puzzle has been compiled, someone else can then solve it. With the general theory of relativity, it is as if nature had set the puzzle, knowing the answers that ought to fit into the squares. What Einstein did was to find out what the pattern of squares was, and to identify the clues that made it possible to solve the puzzle – all without knowing what words were supposed to fit into those squares. Then, armed with Einstein's grid for the pattern of squares and the clues he had uncovered, someone else could solve the puzzle.

The man who did so was a senior astronomer, six years older than Einstein, who was already in his forties when the First World War broke out. But Karl Schwarzschild was a patriot who left his secure post as director of the Potsdam Observatory to volunteer for army service (he had previously, incidentally, been the head of the observatory in Göttingen, yet another connection between that city and the development of relativity theory). Of course, the authorities appreciated that his skills were too valuable to squander in routine soldiering. Once in military service, Schwarzschild worked first in Belgium, at a weather station, and then in France, calculating the trajectories for shells fired over very long range. Then he was sent to the Russian front. It was while serving on the Eastern Front that he contracted a rare skin disease known as pemphiga, which in those days was inevitably fatal.

Schwarzschild, who kept in contact with fellow scientists throughout his military service, learned of Einstein's latest work late in 1915, and was immediately intrigued by it. He had, after all, held the same post in Göttingen that Gauss had held a few decades before him, and Schwarzschild himself was one of those people, following Clifford, who while Einstein was still an undergraduate had speculated that the geometry of space might be non-Euclidean. The papers Einstein presented to the Academy on his behalf were completed only very shortly before Schwarzschild succumbed to his fatal illness. On 16 January 1916, Einstein read to the Academy a paper by Schwarzschild describing the exact mathematical form of the geometry of spacetime around a mass concentrated in a single point; on 24 February, he presented Schwarzschild's description of the spacetime geometry around a spherical mass. On 11 May, Schwarzschild died in hospital in Potsdam, five months short of his forty-third birthday. And with all his achievements as an astronomer and director of two of Germany's great observatories, it is for those two papers, written as a serving officer in wartime conditions and completed in the last months of his life, that Schwarzschild's name is remembered today.

Schwarzschild's initial approach echoes the way in which Newton calculated the gravitational force between the Sun and a planet (or between the Earth and the Moon, or the Earth and an apple) *as if* all the mass of each object is concentrated in a point at its centre. As far as anybody outside the object is concerned, this is a perfectly good way to describe the gravitational influence of any mass. But Schwarzschild's solution to Einstein's equations showed that there is no outside for a genuine point mass! *Any* mass concentrated into a mathematical point will distort spacetime so much that space closes up around the mass and pinches it off from the rest of the Universe.

The pinching off happens at a distance from the point which depends only on the mass involved.

Of course, this is unrealistic. Real masses are never actually concentrated at mathematical points. But Schwarzschild went on to show that for any particular mass, there is a real physical significance for the crucial radius, now known as the Schwarzschild radius (or sometimes as the gravitational radius), at which this pinching off occurs. If the appropriate amount of matter gets squeezed into a sphere smaller than the corresponding Schwarzschild radius, even if it is not actually squeezed into a mathematical point, space (not just spacetime) will indeed be curved so much that the mass will be cut off from the Universe outside. Nothing at all, not even light, will be able to escape. The bigger the mass, the bigger the appropriate Schwarzschild radius. For the Sun, it is 2.9 km; for the Earth, 0.88 cm. A typical galaxy has a Schwarzschild radius of a thousand billion kilometres; but even a mass as small as that of a proton has its own Schwarzschild radius, albeit a mere 2.4×10^{-52} cm. In each case, if you could squeeze the appropriate mass within the corresponding radius you would create what is now known as a black hole.

The extreme distortion of spacetime that makes the mass pinch off from our Universe *only* occurs if the mass is squeezed within the appropriate volume. There is no sense in which there is a black hole with a radius of 0.88 cm now sitting at the centre of the Earth, and if you could drill down to the centre of the Earth you would find nothing unusual going on at that distance from the centre – nothing to indicate that this is where the surface of the black hole would form *if* the Earth were squeezed into such a tiny volume. Once such a squeezing had been achieved, however, anything that crossed the spherical surface with the Schwarzschild radius could never escape from the hole in space. Then, the sphere defined by the Schwarzschild radius defines the surface of a black hole, the surface from which the escape velocity is the speed of light.

The distortion of spacetime geometry that produces this effect is best visualized by imagining a curved two-dimensional surface embedded in three dimensions. The geometry of spacetime described by Schwarzschild's solution is exactly equivalent to the shape you would get by rotating a parabola in ordinary space. A parabola is a very simple curve (Figure 2.10), which is made up of a set of points that are each equally distant from a point known as the focus of the parabola and a straight line known as the directrix. If you

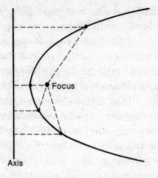

Figure 2.10 *Any point on the curve known as a parabola is the same distance from a point known as the focus and a straight line known as the directrix or axis.*

imagine twiddling the parabola around the directrix, you will get a smoothly curved surface with a wide throat that narrows down into a waist (Figure 2.11). Far away from the waist, the curved surface

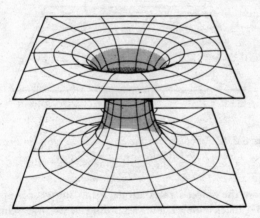

Figure 2.11 *The curvature of space around a black hole echoes the shape you would get by twirling a parabola around its axis.*

flares out and becomes flat – equivalent to saying that the gravitational force is very weak. The more the surface is curved, the stronger the gravitational force at that point, and therefore the greater (in familiar Newtonian language) the escape velocity. If you imagine sliding in along the flared surface towards the waist, it will be harder and harder to escape as the parabolic walls get steeper and steeper. At some critical distance from the waist, which you could mark by drawing a circle around the throat of the surface, it will be impossible to escape. That circle drawn around the throat of the hole is the equivalent on the two-dimensional curved surface to the sphere which marks the surface of a black hole in the three-dimensional curved space of our Universe. Anything that crosses that line can never get out again.

Far away from any mass, the spacetime curvature is the same, whether or not the mass is squeezed within its Schwarzschild radius. Schwarzschild's solution to Einstein's equations then gives exactly the same rule of gravity as Newton did – the inverse square law. But if (like the Sun) the mass is actually *not* sufficiently concentrated to make a black hole, then instead of the parabolic surface forming a waist it has a rounded bottom, like a well (Figure 2.12); the sides never get steep enough to make it impossible to

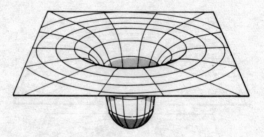

Figure 2.12 *An object like the Sun makes a 'dent' in space like that of a black hole but with a rounded bottom.*

escape from the object's gravitational grip. But what happens at the centre of a black hole? There, according to Schwarzschild's solution, the curvature becomes *infinite*. Such an infinity is known as a singularity. Equations that imply the existence of singularities are usually regarded as flawed – infinities indicate to most physicists that something is wrong with our reasoning, not with the way the Universe works. Partly because of this, Schwarzschild's singular solution to Einstein's equations was not taken at face value by physicists for many years. Although some mathematicians tinkered with the equations, and even though the general theory of relativity passed every other test with flying colours, Schwarzschild's solution was not regarded as having any practical significance in the real Universe. And yet, there the equations were – a complete, accurate representation of the spacetime geometry of a spherical black hole, worked out before the ink was even dry on Einstein's papers presenting the full version of the general theory of relativity itself, and published in 1916, three years before the epic test of light bending during an eclipse that established the accuracy of Einstein's theory to the world at large. It was fully fifty years after Schwarzschild's death that the genuine physical significance of his singular solution as a description of real objects in our Universe began to dawn on astronomers.

CHAPTER THREE

Dense Stars

Degenerate dwarfs that burn white hot, and Indian insights into the fate of matter. Beyond the quantum limit. Stardeath. Black holes discovered again – and forgotten for twenty-five years! Scruffy pulsars, little green men, and confirmation from the Crab

It is straightforward to calculate the distances to the planets of the Solar System, and to the Sun itself, using what is essentially an application on the grand scale of the triangulation techniques used by surveyors here on Earth. That is, the *calculations* are simple, once the painstaking measurements required have been carried out. Such measurements involve, for example, observing the position of Mars against the backdrop of distant stars simultaneously from opposite sides of the Atlantic Ocean, and using the observations to calculate the geometry of a long, thin triangle, with a base stretching from Europe to the Americas, and Mars at its tip. With the distance to the Sun known, its actual size can be worked out from the apparent size of the solar disk on the sky – it is about 109 times bigger across than the Earth is, which means that more than a million spheres the size of the Earth would fit inside the Sun.* And if we know the distance to the Sun, and how long it takes the Earth to complete one orbit around the Sun (one year), we know how big the force of attraction holding the Earth in its orbit is – which, once

* The volume of a sphere is proportional to the cube of its radius, and 100^3 is 1,000,000.

we know the constant of gravity, G, tells us what the mass of the Sun is.

The mass of the Sun is about one third of a million times the mass of the Earth. Since the volume of the Sun is about a million times that of the Earth, this means that the average density of the Sun is just one third that of the Earth. So the density of the Sun is only about one and a half times the density of water, because the average density of the Earth is about 4.5 times the density of water.

It may seem surprising that the density of a star is so low, and in particular that it is less than the density of a typical planet. But remember that this is an *average* density. The simple average conceals the fact that there is an enormous variation of density within the Sun, from a tenuous atmosphere of thin gas to a core where the density is many times greater than that of lead (although, astonishingly, the conditions of pressure and temperature are so extreme at the heart of the Sun that the material of the core still behaves like a gas). The density variations, and the overall average, are very much in line with the structure inferred from calculations of the physical processes which keep stars hot.★

The temperature at the surface of a star is directly related to its colour – blue-white stars are hotter than yellow stars, yellow stars are hotter than red stars, with a whole range of subtle variations in between (our Sun, a yellowish-orange star, with a surface temperature of just under 6,000°C, is pretty much in the middle of the range of stellar colours). You might expect that, by and large, hotter stars are brighter than cooler stars – and, by and large, you would be right. But there are exceptions. This rule of thumb applies only if stars are more or less the same size as each other. One of the simplest features of stars in general is that their brightness depends both on how hot they are and on how big they are. A star which has a large surface area, radiating energy outward across every square metre of its surface, can have a fairly cool surface and still be very bright, simply because there are so many square metres doing the radiating. In order to shine with the same brightness overall, a smaller star would have to have a hotter surface, so that each square metre could radiate more energy than the same area of the larger star. And at just about the same time as Einstein was completing his general theory of relativity, and Schwarzschild was

★ The physics of the Sun in particular, and stars in general, is described in my book *Blinded by the Light*.

using Einstein's equations to describe the structure of a black hole in mathematical terms, observational astronomers were puzzling over the discovery that some hot stars are also very dim, and seemed to be not much larger than the Earth, even though they contained nearly as much mass as our Sun. Taken at face value, this seemed to imply that the *average* density of such a star must be about a hundred *thousand* times the density of water.

Dwarfish companions

In fact, the first hint that such dense stars might exist came in the 1840s. It resulted from observations made by the German astronomer Friedrich Bessel, very late on in his career. Bessel had been born in 1784, and died in 1846, only a couple of years after finding the first evidence that small, dense stars exist. He is best-known, though, not for this discovery but for his greatest achievement, which was to measure the distance to another star. He did this by extending the surveyor's triangulation technique to its limit. Instead of observing Mars from opposite sides of the Atlantic, Bessel observed a star known as 61 Cygni when the Earth was on opposite sides of its orbit around the Sun – the observations were made six months apart. This gave him a base line 300 million km long (the diameter of the Earth's orbit), and showed up a tiny apparent movement of 61 Cygni against the background stars (the effect, known as parallax, is actually due to the movement of the Earth around its orbit). Bessel worked out that the star was so far away from us that light must take several years, at a speed of 300,000 km a second, to travel across space from 61 Cygni to the Earth – it is several light years away. This was the beginning of an understanding of the true scale of the Universe.

As part of the background to this work, Bessel had measured and catalogued the accurate positions of some 50,000 stars. Only the nearest stars are close enough to us for their distance to be measured by the parallax effect; for most stars, the distances are so huge that even a baseline 300 million km across is not enough to produce a measurable parallax. The discovery of dense stars stemmed from Bessel's observation that some stars really do move in a rhythmic way against the background sky, not because of the optical illusion of parallax, but because something is tugging on them. He found that two bright stars, Sirius (which is actually the brightest star in the night sky, partly because it really is bright but also because it is relatively close to us) and Procyon, both showed such a rhythmic, regular motion, wiggling from side to side. This could not be explained as a parallax effect caused by the motion of the Earth around the Sun. But it could be explained as the result of a force, in each case, tugging the visible star from side to side. The natural explanation for such a rhythmic tug acting on these stars was that each of them had a companion, an unseen star in orbit around them, pulling on them by gravity.

Careful observations of the way in which Sirius moved showed what kind of orbit the unseen companion must be in. Sirius is twice as bright as any other star in the night sky, and is easily visible near the constellation Orion. Because it is so close to us (only 8.7 light years away), its motion through space shows up against the background of the 'fixed' stars, which are actually all moving, but are so remote that they do not seem to move at all from one year to the next. Even the motion of Sirius across the sky amounts to a mere 1.3 seconds of arc each year – 0.07 per cent of the distance across the full Moon as seen from Earth. It is the fact that this motion across the sky is not in a straight line, but has a tiny wiggle, that reveals the presence of a companion star. The wiggle tells us that the companion orbits Sirius once every 49 years, and, armed with Kepler's laws of orbital motion and Newton's law of gravity, astronomers could then calculate the masses of both Sirius and its companion. Sirius weighs in at just under two and a half times the mass of our Sun, while its companion, now known as Sirius B, has a mass roughly 80 per cent of that of the Sun. Sirius is a hot, white star; its unseen companion, it seemed obvious in the middle of the nineteenth century, must be a cool, dim star.

The first person to see the companion of Sirius was Alvin Clark, an American telescope maker. In 1862, while testing a telescope with a new 18-inch lens, which was destined for installation in the Dearborn Observatory in Illinois, he turned the instrument on to Sirius. The telescope was so good that he could see the companion, which turned out to be so faint that if it were the same distance from us that the Sun is, it would shine only one four-hundredth as brightly in the sky. The faintness of the companions to Sirius and Procyon puzzled astronomers for another fifty years, and in the early years of the twentieth century the puzzle was compounded by the discovery of one or two similar celestial objects. The initial response of astronomers to the puzzle was accurately summed up in a book by the American astronomer Simon Newcomb, published in 1908. Referring to the companions of Sirius and Procyon, Newcomb said that 'either they have a far less surface brilliancy than the sun or their density is much greater. There can be no doubt that the former is the case.'* The two alternatives that Newcomb presented were, indeed, the only possible solutions to the puzzle; but his conclusion was wrong.

Even Newcomb must have had some doubts about the conclusion, for although it is difficult to see Sirius B clearly in a telescope because of the glare from Sirius itself, as far as anyone could tell the companion star seemed to be a white (which meant hot) star, the same colour as Sirius itself. This suspicion was confirmed six years after Newcomb died. In 1915, Sirius B was at its furthest distance from Sirius in its orbit, so it could be seen relatively clearly. In December that year Walter Adams (an American astronomer who had actually been born, in 1876, in Syria, where his parents worked as missionaries) obtained the first spectrum of Sirius B. In this connection, the important feature of such a spectrum is that it shows how much energy the star is radiating at different wavelengths (different colour bands), and gives a precise measure of its temperature, as well as its colour. The spectrum of the companion turned out to be identical to that of Sirius. It was the same colour as Sirius, which meant that its surface was at the same temperature as Sirius, which (in order for it to be so faint overall) meant that it was much smaller than Sirius, only a little bigger than the Earth.

The only other alternative, which a few astronomers tried to cling to for a few months, was that the companion did not shine by its own light at all, but merely reflected the white light from Sirius, in the way that the Moon reflects the light from the Sun. But Adams had an answer to that: he pointed out that another star, Eridani B, also had a very low overall brightness, but its spectrum was like that of Sirius B, even though, in this case, there was no white companion star whose light might be being reflected. Eridani B (and, by implication, Sirius B) had to be both white and small – a

---

* *The Stars*, published by John Murray, London. Newcomb (who was born in Canada) lived from 1835 to 1909, and had a distinguished career as a scientific officer in the US Navy, working at the Naval Observatory in Washington and at Johns Hopkins University, as well as being a founder of the American Astronomical Society and its first President. His comment is, indeed, representative of the Establishment view at the time.

white dwarf star, with a density, in round terms, ten thousand times the density of lead.

Strangely, the oddity of Eridani B had been noticed in passing five years earlier as a result of a chance remark by an astronomer at Harvard. But none of the three people who spotted the oddity followed their observation through. The astronomer who made the chance observation was Henry Norris Russell, later to become one of the co-inventors of a method for relating the brightness of a star to its temperature (or colour) on a kind of graph known as a Hertzsprung-Russell diagram. The famous diagram emerged in 1913; but the work which led up to it, involving studies of the colours of stars with different brightnesses, was already well under way in 1910.

Stars are classified by colour in terms of a system worked out at the Harvard College Observatory in the early 1900s. What should have been a simple alphabetical list got shuffled up, because some of the alphabetical labels were attached to the wrong kinds of stars early on, before their actual properties were known accurately. So, to this day, stars are classified by colour in terms of the labels O, B, A, F, G, K, M. O and B stars are white and hot; K and M stars are cool and red. Our Sun is an orange G-type star.* Russell needed to know the spectra of as many stars as possible in order to find general rules relating colour and brightness, and Edward Pickering, the director of Harvard College Observatory, had agreed to provide spectra for stars that had been observed during parallax studies of the kind pioneered by Bessel. This led Russell to the discovery that all the very faint stars on the list were of type M.

Many years later, Russell recalled ruefully how he had been discussing this discovery with Pickering one day in 1910, and mentioned that it would be interesting to check whether other faint stars fitted the pattern.

> Pickering said, 'Well, name one of these stars.' 'Well,' said I, 'for example the faint companion of Omicron Eridani.' So Pickering said, 'Well, we make rather a speciality of being able to answer questions like that.' And so we telephoned down to the office of Mrs Fleming and Mrs Fleming said, yes, she'd look it up. In half an hour she came up and said, 'I've got it here, unquestionably type A.' I knew enough, even then, to know what that meant. I

---

* The Harvard classification system is made memorable by a mnemonic invented in those days of unthinking male chauvinism, which runs 'Oh, Be A Fine Girl, Kiss Me'.

was flabbergasted. I was really baffled trying to make out what it meant. Then Pickering thought for a moment and then said with a kindly smile, 'I wouldn't worry. It's just these things which we can't explain that lead to advances in our knowledge.' Well, at that moment Pickering, Mrs Fleming and I were the only people in the world who knew of the existence of white dwarfs.*

Pickering was right. It was indeed the failure of white dwarfs to fit the pattern that was to lead to further advances in knowledge. Even after Adams obtained the spectrum of Sirius B, however, it was to take almost another twenty years for the puzzle to be solved – and even then, not every astronomer was happy with the solution.

Degenerate stars

An understanding of the nature of white dwarf stars emerged during the 1920s, as part of the developing understanding of the internal structure of stars in general. This work was pioneered by the same Arthur Eddington who had been involved in measuring the light-bending effect in 1919, and who, as an expert on both general relativity and stellar structure, was a metaphorical giant who towered over the astronomical community of the time.

The understanding of how stars work developed only slowly because it required an understanding of how atoms work, which was itself only developed, in the form of quantum mechanics, in the middle of the 1920s. One of the key ingredients in the new understanding of stellar structure was the realization which explained why the deep interior of a star like the Sun can be described as if it were a perfect gas, even though it has such a very high density. The secret lies in the fact that an atom is made up of a tiny nucleus, composed of particles called protons and neutrons, surrounded by a cloud of smaller particles known as electrons. The size of the nucleus, in relation to the size of the whole atom, is like a grain of dust in the middle of a football stadium. The nucleus actually takes up only about one trillionth of the volume of the atom.

In a gas, such as the air that you breathe, atoms are moving about

* From a talk by Russell at Princeton University Observatory in 1954 (three years before he died, in his eightieth year); referred to in the *Source Book* edited by Kenneth Lang and Owen Gingerich (see the Bibliography).

rapidly, constantly bashing against one another, just like tiny spheres of perfectly elastic material. In a solid, the atoms stay more or less in the same place, vibrating gently and jostling against one another. In a liquid, they have just enough energy to slide past one another. In each case, the nuclei of the atoms take no part in the bashing, jostling or sliding – only the electrons, on the outside of the atoms, ever come into contact with one another.

Under the conditions of intense heat and pressure inside a star, however, the collisions are so violent that electrons are knocked off the atoms. This can leave bare nuclei behind, mingling with electrons and with each other in a kind of hot fluid known to physicists as a plasma. If all the electrons are removed from the nuclei, the plasma can be squeezed to one trillionth of the volume of the equivalent cloud of gas, and still behave exactly like a gas – only now, instead of atoms moving at high speed and bouncing off one another, it is nuclei that move at high speed and bounce off one another. This is what happens in the heart of the Sun and most stars. In the process, some of the nuclei collide with one another so hard that they stick together, converting nuclei of hydrogen into nuclei of helium and releasing energy as they do so, which is known as nuclear fusion. That is how stars stay hot.*

But what happens when all the potential for nuclear fusion is exhausted, and a star begins to cool down in its heart? You might expect that the nuclei would recapture their electrons, turning back into atoms and turning the plasma back into a cloud of gas. But in order to do this, they would have to find energy from somewhere, to make the star core expand again and make room for the atoms, in spite of the weight of gravity tugging it inward. Since there is nowhere for the star to get this energy from, it cannot happen. As Eddington used to express it, such a star would have to *gain* energy in order to cool down! 'It would seem', he said in his classic book *The Internal Constitution of the Stars*, 'that the star will be in an awkward predicament when its supply of sub-atomic energy ultimately fails.'

In 1926, the same year that Eddington's book was published, Ralph Fowler, working at the University of Cambridge, showed how a dying star might begin to resolve that predicament. He calculated that according to the new quantum theory what would

---

* I describe how stars stay hot in more detail in my book *Blinded by the Light*; details of quantum physics are explained in *In Search of Schrödinger's Cat*.

actually happen to such a star would be for it to settle down into a very dense state, with atomic nuclei embedded in a sea of electrons. The pressure of the electrons themselves bashing against each other and against the nuclei would balance the inward tug of gravity once the star had shrunk to a certain size. The actual size at which a dense star would stabilize depends on its mass; Fowler calculated the range of possibilities and found that they were very close to the actual masses of white dwarf stars, such as Sirius B. Quantum theory had satisfactorily explained the structure of white dwarf stars; in the language of modern physics, matter under these extreme conditions is said to be 'degenerate', and they are held up by the 'degenerate pressure' of electrons in their lowest quantum energy state – as a 'degenerate electron gas'. Before the decade was out, however, a few astrophysicists began to realize that when the effects of relativity are taken into account, as well as those of quantum mechanics, even a degenerate electron gas cannot hold all dense stars up against the inward pull of gravity. And this is a consequence *not* of the general theory of relativity, but of Einstein's older special theory.

The white dwarf limit

The set of rules that physicists use to describe the properties of something like a gas, or a plasma, is known as an equation of state. It enables you to calculate the way the gas changes when the conditions it experiences change – what happens to the volume, for example, if you double the pressure. The density of the material at the heart of a star depends on how much mass the star contains, because the more mass there is, the harder is the gravitational squeeze; a good equation of state will tell you what central density corresponds to a particular stellar mass, taking account of all the physics going on inside the star. By 1929, Edmund Stoner, of Leeds University, had shown that even allowing for quantum effects there must be a maximum density for the material in a degenerate star, when all of the electrons are, in effect, wedged together as tightly as possible. The density he came up with was about ten times greater than the density of known white dwarfs, so at first sight this didn't look too worrying. But almost immediately Wilhelm Anderson, of Tartu University in Estonia, pointed out that under the extreme conditions described by Stoner the electrons inside such a star would have to bash against one another so hard, in order to hold the star up against the inward tug of gravity, that they would have to be moving at close to the speed of light, even though they would each only move a short distance before colliding with a nucleus, rebounding, and colliding again in a never-ending dance like the ball in some frantic cosmic pinball machine. If such high speeds are involved, the equation of state has to take account of the effects predicted by the special theory of relativity, and in particular of the way in which the mass of each electron will increase as it goes faster. This meant that the maximum possible density for a white dwarf could not be very much greater than that of Sirius B, after all. Taking up this point, Stoner came back into the fray, developing what became known as the Stoner-Anderson equation of state, and showing in 1930 that when more accurate account was taken of relativistic effects even quantum effects could not stabilize any degenerate star with a mass more than 1.7 times that of our Sun. But he commented only that all known white dwarfs did indeed have masses below this limit, and did not speculate on what might happen to stars with more mass, once they had run out of nuclear fuel.

In fact, the limiting mass worked out by Stoner was only approximate. He had not put all of the astrophysical details into his calculation – for example, he had treated the 'star' in his equations as if it had the same density all the way through, instead of being more dense in its heart. The person who put this kind of calculation on a more accurate basis, and who correctly worked out that the actual limiting mass for a white dwarf star made of helium is a little over 1.4 solar masses, was a remarkable Indian scientist. He made the calculation, completely unaware of the work by Stoner and Anderson, to pass the time while travelling by boat from India to England to carry out research as a student in Cambridge. And he was only nineteen years old at the time.

Subrahmanyan Chandrasekhar was born in Lahore (then part of British India, now in Pakistan) on 19 October 1910 (the year in which Russell, Pickering and Mrs Fleming accidentally became the first three people to learn of the existence of white dwarf stars), and probably ranks with Eddington as one of the two great astrophysicists of the twentieth century. He received the Nobel Prize in physics in 1983, and the citation for that award refers, among other things, to the calculations he had made during that boat trip in July

1930, more than half a century before. Ironically, however, Chandrasekhar's studies of the structure of white dwarf stars brought him into conflict with Eddington, who never accepted the implications. Indeed, he opposed the idea of a mass limit for stable stars so strongly that, through his influence as an *éminence grise* of the astronomical community, he may have held back the investigation of black holes for a decade or more. This is all the more curious since, as an expert on both the general theory of relativity and stellar structure, Eddington might seem, with hindsight, to have been the ideal person to pick up on the notion. But by the time Chandrasekhar arrived in Cambridge Eddington was only a few months short of his forty-eighth birthday; his greatest scientific achievements were behind him, and he was set in his scientific habits, reluctant to accommodate dramatic new ideas.★

Chandrasekhar, however, virtually cut his teeth on new ideas. He was a student in Madras just at the time when the new quantum theory was being developed in Europe, and as well as text-books such as Eddington's own *The Internal Constitution of the Stars* and Arnold Sommerfeld's *Atomic Structure and Spectral Lines* he read the scientific papers of quantum pioneers such as Niels Bohr, Werner Heisenberg and Erwin Schrödinger in the research journals in the college library. We know a great deal about this period in Chandrasekhar's life thanks to an interview he gave in 1977, which is preserved in the Niels Bohr Library of the American Institute of Physics. 'I wasn't taught quantum mechanics,' he said. 'I learned it from Sommerfeld's *Atomic Structure and Spectral Lines*.' Indeed, all the evidence is that even before he graduated in 1930, Chandrasekhar knew more about physics than the teachers at his college did. He actually published two research papers while still an undergraduate, and on the strength of this work won a scholarship to study in England. So his calculations on the trip over hardly came out of the blue.

In Cambridge, Chandrasekhar officially came under the wing of Ralph Fowler, who was the supervisor for his doctoral studies (in fact, he was almost completely ignored by Fowler, whom he saw only once in six months, and worked largely on his own: this is not untypical of the way Cambridge treated research students). Proudly Chandrasekhar showed Fowler the calculation which said that

★ None of this seems to have caused any bitterness in Chandrasekhar, however, who as a teenager virtually hero-worshipped Eddington, and who much later wrote a sympathetic biographical memoir of him (see the Bibliography).

white dwarfs had to have masses less than 1.4 times that of the Sun – but Fowler, in spite of his own earlier work on degenerate stars, didn't seem to think it was important. 'I didn't understand at the time what this limit meant,' Chandrasekhar recalled in 1977, 'and I didn't know how it would end. But it is very curious that Fowler did not think the result very important.' Nevertheless, Chandrasekhar's calculations were published, in 1931, in the *Astrophysical Journal* – a nice historical touch, since he became the editor of that very journal in 1953, and ran it until 1971. And throughout the early 1930s he persevered, with little encouragement, in his efforts to find out what the white dwarf limit really meant.

You can get an accurate impression of how scientists responded to the suggestion that there must be an upper mass limit for white dwarfs from a comment made by the Soviet physicist Lev Landau, in a paper published in 1932. Landau did not know about Chandrasekhar's work, and worked out the same limiting mass quite independently. He made one howler in his paper – not being an astronomer, he omitted to take proper account of the role of ordinary gas pressure in holding stars up against gravitational collapse as long as they have nuclear fuel to burn in their centres. But he did work out the correct mass limit for degenerate stars, and said that for any star with more mass than this 'there exists in the whole quantum theory no cause preventing the system from collapsing to a point.'* The quantum theory was still less than ten years old at that time, and Landau felt no qualms about what would have to give when push came to shove. If quantum theory said that stars with more than about one and a half times the mass of the Sun could not be held up even by the pressure of a degenerate electron gas, quantum theory must be wrong – 'we must conclude that all stars heavier than 1.5 solar masses certainly possess regions in which the laws of quantum mechanics . . . are violated.'

But Chandrasekhar was busily closing down loopholes in the calculation. He completed his Ph.D. studies in 1933, at the ripe old age of twenty-two, and was elected a Fellow of Trinity College. With the confidence of his new status, Chandrasekhar worked during 1934 on a presentation of his completed theory for white dwarf stars, which he gave as a talk to the Royal Astronomical Society in London in January 1935. Immediately after the talk, Eddington got up and said that Chandrasekhar's theory was

* Quoted by Werner Israel, in *300 Years of Gravitation*, edited by Hawking and Israel.

complete rubbish. But his opposition to the idea of a limiting mass
was no more based on physics than Landau's dismissal of his own
calculations. Like Landau, Eddington relied on commonsense to
tell him where the laws of physics could or could not be applied.
But in his own presentation to that meeting of the Royal Astronomical Society, Eddington came tantalizingly close to realizing that
black holes with masses comparable to that of the Sun must exist.
He said:

> Chandrasekhar, using the relativistic formula which has been
> accepted for the last five years, shows that a star of mass greater
> than a certain limit M remains a perfect gas and can never cool
> down. The star has to go on radiating and radiating, and
> contracting and contracting, until, I suppose, it gets down to a
> few km radius, when gravity becomes strong enough to hold in
> the radiation, and the star can at last find peace.

Had he stopped there, Eddington would now be remembered as the
father of black hole astrophysics. Alas, he had mentioned the
possibility of gravity distorting spacetime so much that it would
trap light only to poke fun at Chandrasekhar. Scarcely pausing to
draw breath, Eddington went on:

> Dr Chandrasekhar had got this result before, but he has rubbed
> it in in his last paper; and, when discussing it with him, I felt
> driven to the conclusion that this was almost a *reductio ad
> absurdum* of the relativistic degeneracy formula. Various acci-
> dents may intervene to save the star, but I want more protection
> than that. I think there should be a law of nature to prevent a star
> from behaving in this way!*

Eddington's objections continued over the next few years, but he
was never able to find a law of nature that would save an
overweight white dwarf from collapse. Astrophysicists were left
with only the possibility of the 'various accidents' Eddington
referred to, which might make a massive star lose material as it
aged, blowing it away into space so that whatever mass it started
out with it must end its life with less than what became known as
the Chandrasekhar limit. Even Chandrasekhar himself speculated
along these lines, and the option of such accidents conspiring to get

---

* Originally published in *The Observatory*, Volume 58, p. 37, 1935; quoted by Chandrasekhar in his book *Eddington*.

rid of the excess mass was still being taught as a viable possibility in the early 1960s (indeed, it was an option presented in all seriousness by my own lecturers when I was a student, as late as 1966). But this never seemed a very plausible scenario – how, after all, could a star that started out with, say, ten times as much mass as our Sun 'know' just how much gas it had to puff away into space during its lifetime in order to end its days as a stable white dwarf? The only reason anyone ever took the idea even half seriously was because they simply could not bring themselves to accept the only alternative, that some stars really must end their days in an ultimate gravitational collapse.

It took a long time for Chandrasekhar's ideas about the structure of white dwarf stars to be fully accepted, although his mass limit takes its place in all the text-books published after 1936. Looking back from 1977 to his dramatic run-in, as a young researcher in his twenties, with the grand old man of astrophysics, Chandrasekhar said, 'I am astonished that I was never completely crushed.' But he was partly crushed by the onslaught. He left Trinity College in 1936 to work at the University of Chicago, and 'finally, in 1938, I decided that there was no good in my fighting all the time, claiming that I was right and that the others were all wrong. I would write a book. I would state my views. And then I would leave the subject.'

The book, *An Introduction to the Study of Stellar Structure*, was published in 1939, and like Eddington's *The Internal Constitution of the Stars* (published in 1926) it became a classic, still used by students of astrophysics today. True to his word, Chandrasekhar then turned his attention to other problems, setting a pattern which persisted throughout his working life. He would spend several years working in a particular area, then write a comprehensive book on the subject before moving on again to pastures new. This career pattern led him, through the study of stellar dynamics, stellar atmospheres and other research topics to major work on the application of the general theory of relativity to astrophysics in the 1960s, and to work on the mathematical theory of black holes in the 1970s and 1980s. The wheel had turned full circle, so that when Chandrasekhar received the Nobel Prize for his work in relativity and black hole investigations, it recognized both his latest work and the work with which he had made his name half a century before. After the mid-1930s, though, the investigation of the physics of dense stars was left in other hands. It turned out that the white dwarf limit was not, after all, quite the end of the story of degenerate stars. There is, in fact, one more stepping stone on which a dead star can rest without suffering the fate, so scornfully dismissed by Eddington, of 'radiating and radiating, and contracting and contracting until . . . gravity becomes strong enough to hold in the radiation.'

The ultimate density of matter

Although I have described the structure of an atom in terms of protons, neutrons and electrons, in 1930, when Chandrasekhar first worked out his famous mass limit, nobody knew that neutrons existed. The only particles known to physicists were the electrons, which each carry one unit of negative electric charge, and the protons, which are much more massive than electrons and each carry one unit of positive charge. Early descriptions of the degenerate material of which white dwarfs are made referred only to atomic nuclei and electrons, because it was not clear at that time exactly what nuclei were made of. Things changed in February 1932, when James Chadwick, working at the Cavendish Laboratory in Cambridge, identified the neutron. This is a particle with almost the same mass as the proton, but with zero electric charge – the first electrically neutral particle to be discovered. Once the neutron had been found, it was natural that some physicists and astronomers should begin to speculate about the possible existence of stars made entirely or partly of neutrons, and to wonder whether, in the light of Chandrasekhar's curious result, there was any upper mass limit to the stability of such stars.

Probably the first physicist to make any calculations along these lines was Lev Landau. In his original work on degenerate stars, he had talked of the possibility that all stars might contain a core of degenerate nuclear material, kept stable even for massive stars (in spite of what the rules of quantum physics said) by some unknown means. Landau was nearing the end of a visit to Niels Bohr's research institute in Copenhagen when news of Chadwick's discovery came through, and according to other researchers who were present at the time he immediately began to talk about the possibility of stellar cores made purely of neutrons. But he returned to the Soviet Union later in 1932, and didn't publish any of his ideas along these lines until 1938. Meanwhile, though, news of Landau's speculations had been brought out of Russia and publicized by

George Gamow, a Ukrainian astrophysicist who found Stalin's regime uncongenial and fled to the West in 1933.

The speculation that stellar cores might consist of dense masses of neutrons was attractive to astrophysicists at the time, because even in the middle of the 1930s they did not know how stars kept themselves hot inside. The favoured idea was that some form of nuclear fusion reactions provided the energy to keep stars like the Sun hot for billions of years; but since nobody had worked out exactly which nuclear reactions might do the job under the conditions of temperature and pressure at the heart of a star, the way was still open for discussion of alternative ideas. The neutron core idea suggested that such a core at the heart of a star might grow very slowly as successive layers of ordinary material around the core collapsed into the ball of neutrons. This steady shrinking of the outer part of the star on to the degenerate core would slowly release gravitational energy, which would appear in the form of heat. In order to generate the amount of energy radiated by the Sun in a billion years, Landau said, just 1 per cent of the material inside the Sun would have to collapse in this way.

There was even a suggestion as to how the collapse would occur. Soon after the neutron was discovered, physicists found that a neutron left on its own, not in an atomic nucleus, lasts, on average, for only a few minutes. It soon 'decays', spitting out an electron and turning into a proton. The process is known as beta decay. The opposite process can also happen – a fast-moving electron can penetrate a proton, combining with it to become a neutron. This is known as inverse beta decay. What could be more natural, thought Gamow, Landau and a few others, than that, under the conditions of high pressure and temperature in the core of a star, electrons might be steadily forced to combine with protons to create neutrons, adding to a growing ball of neutron stuff like a single huge atomic nucleus at the star's centre?

The rug was pulled from under all such speculations, however, when physicists identified, at the end of the 1930s, the set of nuclear reactions that actually does convert hydrogen into helium inside a star like the Sun, keeping it hot by nuclear fusion. The calculations fitted so beautifully with observations of the properties of stars that no room was left for continued speculation about growing neutron cores. And the last nail in the coffin of Landau's intriguing idea was hammered home by Gamow and one of his colleagues, M. Schönberg, when they showed, in 1941, that 'neutronization' of a stellar core, if it ever did begin to happen, would be an all-or-nothing, runaway process in which the whole mass of the inner part of the star would suddenly collapse inward into a ball of neutrons, releasing a vast amount of gravitational energy in one mighty explosion. That, however, was good news for one other astronomer, who had suggested seven years previously that neutron stars might be formed during the great stellar explosions known as supernovas. In spite of the calculation by Gamow and Schönberg, though, it was to take more than thirty years for the rest of the astronomical community to come round to his way of thinking.

The man who was so far ahead of the pack was Fritz Zwicky, who had been born in Bulgaria (in 1898) but whose parents were Swiss, and who remained a Swiss national throughout his life, even though he worked in California from 1925 onwards. Zwicky died only in 1974, so he at least had the satisfaction of living to see his ideas about supernovas accepted, even if he did have to wait thirty years.

Supernova explosions are the biggest stellar explosions that occur in the Universe today. Although they are rare, when they do occur a single star releases so much energy that for a brief period it will shine as brightly as a whole galaxy of stars like our Milky Way – even though a galaxy contains typically a hundred billion ordinary stars. In 1934, in a paper that he wrote with the German-born astronomer Walter Baade, who had emigrated to America in 1931, Zwicky pointed out that such an enormous outburst of energy must involve the conversion of an appreciable fraction of the dying star's mass into pure energy, in line with the prediction of the special theory of relativity that matter and energy are interconvertible. The same year, Baade and Zwicky published another paper, mainly concerned with the notion that particles known as cosmic rays, which arrive at the Earth from space, are produced in supernova explosions. At the end of that paper, in a kind of afterthought that was chiefly Zwicky's idea (and which was really an afterthought related to their *previous* supernova paper), they commented:

> With all reserve we advance the view that a super-nova represents the transition of an ordinary star into a *neutron star*, consisting mainly of neutrons. Such a star may possess a very

small radius and an extremely high density . . . A neutron star would therefore represent the most stable configuration of matter as such.★

This proposal, coming just two years after the discovery of the neutron, was a much more daring leap of intuition than it may seem from the perspective of the 1990s. After all, in 1934 astrophysicists had barely come to terms with the idea of white dwarfs. But while a white dwarf is, in terms of its radius, about one hundredth the size of the Sun, a neutron star is only one *seven*-hundredth of the size of a white dwarf! Such a star would contain roughly as much matter as our Sun, but packed into a sphere only about 10 km across. A white dwarf is about two thousand times bigger than the Schwarzschild radius for the amount of matter it contains – far enough removed from the prospect of becoming a black hole to keep physicists reasonably happy about avoiding the prospect of ultimate gravitational collapse. But if neutron stars exist, they are only about three times bigger than their own Schwarzschild radii – far too close for comfortable complacency (Figure 3.1). A neutron star sits on

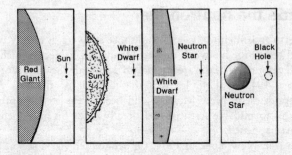

Figure 3.1 *Relative sizes of astronomical objects. In terms of its diameter, a red giant is 200 times bigger than the Sun. The Sun is 100 times bigger than a white dwarf; a white dwarf is 700 times bigger than a neutron star; but a neutron star is only three times bigger than a black hole (the Earth is about the same size as a white dwarf). So when neutron stars were discovered, many astronomers began to believe that black holes must also exist.*

★ Quoted in Lang and Gingerich, *Source Book*.

the very threshold of becoming a black hole. Indeed, if you really believed that neutron stars existed, you might have to accept that black holes existed, as well!

Small wonder that astrophysicists shied away from the prospect, preferring to believe, until well into the 1960s, that even an explosion as violent as a supernova would leave behind nothing more compact than a white dwarf. After all, white dwarfs were known to exist, but nobody had ever seen a neutron star. And wouldn't a supernova explosion be just the thing for getting rid of excess mass and ensuring that the remnant left behind weighed less than the Chandrasekhar limit? It was easy to think so at the end of the 1930s. But before the study of such collapsed objects went into hibernation for a quarter of a century, there was one last flurry of theoretical activity. In the wake of Landau's suggestion that all stars might contain neutron cores, one team of American researchers looked into the question of whether such cores, or even complete neutron stars, could indeed be stable, and if there was a mass limit for them like the Chandrasekhar limit for white dwarfs. The answer, to both questions, was 'yes'.

Inside the neutron star

Robert Oppenheimer, who found those answers, is chiefly remembered today for his work on the Manhattan Project, which led to the development of the atomic bomb during the Second World War – from 1943 to 1945 he was the director of the Los Alamos Laboratories, in New Mexico, and headed the atom bomb team. But this formidable scientist had already made his mark before any of this happened.

Born in New York in 1904, Oppenheimer was a sober, serious child, always top of his class at school. He went to Harvard at the age of eighteen, and graduated *summa cum laude* in 1925, having completed what was officially a four-year course in just three years. He moved on to study in Europe with the pioneers of the new theory of quantum physics, first in Cambridge and then in Göttingen, where he obtained his Ph.D. in 1927. Back in the United States, in 1929 Oppenheimer was appointed to a joint post as Assistant Professor in *both* the California Institute of Technology *and* the University of California at Berkeley. Commuting between the two campuses, he satisfied the demands of this dual role so well that he was promoted to Associate Professor in 1931, and full Professor in 1936.

At first, although Oppenheimer, with his recent experience in Europe, knew more about quantum physics than just about anybody on the West Coast, he was a hopeless teacher, who raced through his lectures too fast, mumbled, and smoked almost non-stop; the story is that students, watching but not comprehending the display as Oppenheimer chalked equations on the board with one hand and held a cigarette in the other, would place bets on whether he would try to write with the cigarette or smoke the chalk. But apparently he never did. What he did do, though, was to learn from the comments of the students where he was letting them down. He slowed down his classroom presentations, made them more clear, and spent a lot of time out of the class with his graduate students, becoming one of the best teachers of physics on either campus in the 1930s. His interests spread widely across the new developments in physics, and it was natural that he should be intrigued by the idea of neutron cores, and should enlist some of those graduate students to work with him on an investigation of their behaviour.

Gamow had published some speculations based on Landau's idea in 1937, and Landau's own thoughts about neutron cores appeared in print in 1938. Landau's hope that slow collapse of a star on to a neutron core might release energy and let the star shine brightly for a long time would only stand up, of course, if neutron cores themselves could stand up against the inward tug of gravity. Landau himself estimated that such a neutron core could be stable if it had a mass less than about 5 per cent of the mass of the Sun, but his calculation was very simplistic, and, among other things, did not take account of the effects of neutrons themselves reaching pressures where they would behave as a degenerate relativistic gas. In 1938 Oppenheimer and his student Robert Serber pointed out a flaw in Landau's own calculation, which when carried through more accurately led to a mass estimate as high as 30 per cent of a solar mass; but that quick response to Landau's paper still did not take account of neutron degeneracy. When Oppenheimer and another student, George Volkoff, tackled this aspect of the puzzle, and also included an allowance for the distortion of spacetime produced by gravity at the enormous densities inside neutron stars, they concluded (in a paper published early in 1939) that stable neutron stars (or cores) could exist only if they had masses in the range from 10 to 70 per cent of the mass of our Sun, corresponding to densities in the range from one hundred thousand billion grams per cubic centimetre to ten *million* billion grams per cubic centimetre. For masses greater than the 'Oppenheimer-Volkoff limit', there was no way to hold a star up even by enlisting the aid of relativistically degenerate neutrons, and they wrote that 'the star will continue to contract indefinitely, never reaching equilibrium.'[*]

Like Eddington, Oppenheimer found such a prospect unpalatable. 'One would hope', the paper with Volkoff continued, that there might be solutions to the equations 'for which the rate of contraction, and in general the time variation, become slower and slower, so that these solutions might be regarded, not as equilibrium solutions, but as quasi-static.' That is, the way Oppenheimer saw out of the dilemma of ultimate gravitational collapse was the prospect that the distortion of spacetime caused by the gravity of the collapsing star would make time run so slowly that for someone in the Universe outside the collapse would seem to take forever. If it takes an infinite time for a star to collapse to a point of infinite density, we don't have to worry about the possibility that such infinitely collapsed objects might be discovered in the real Universe.

Since 1939, although the equation of state for degenerate neutron material has been improved slightly, the basic conclusions reached by Oppenheimer and Volkoff still hold. Today, the best estimate is that a stable neutron star can exist only if it has a mass of more than 10 per cent of the mass of the Sun,[†] and certainly less than three times the mass of the Sun (possibly only if the mass is less than twice that of the Sun). This corresponds to stars with radii between about 9 km and 160 km (at the outside; probably no neutron star has a radius bigger than 100 km).

There is one last adjustment to the equation of state which has not yet been fully worked out, and is still a contentious issue. It is now thought that neutrons themselves are composed of particles known as quarks, and this raises the possibility that in the centre of

[*] *Physical Review*, Volume 55, pp. 374–81.

[†] Although a lighter ball of material might be squeezed into the neutron state in a stellar explosion, it would be too light to maintain the pressure needed for neutron degeneracy by its own gravity. Once the explosion had faded away, many of the neutrons would turn into protons by beta decay, releasing electrons, and the ball of star stuff would simply become a small white dwarf.

a neutron star these quarks may roam freely in a (relativistically degenerate) fluid form known as a 'quark soup'. But since quarks are, in everyday terms, already 'touching' each other inside a neutron, this possibility does not allow for densities much greater than those of 'ordinary' degenerate neutron material. Even allowing for the presence of quarks, it is still a reliable rule of thumb to say that no stable neutron star can exist with a mass greater than three solar masses.

Beyond the neutron star

Unlike Eddington contemplating the fate of massive white dwarfs, or Landau thinking about neutron cores, Oppenheimer was not prepared to leave it to unknown laws of nature and new forces to stabilize neutron stars heavier than the Oppenheimer–Volkoff limit. And when he found that the rigorous application of general relativity to the problem left no loophole preventing the collapse, he accepted what the equations of the general theory had to say. In July 1939, working now with yet another of his students, a mathematical whizz called Hartland Snyder, Oppenheimer completed a paper that went beyond the investigation of stable neutron stars and looked at the way in which gravity would indeed distort spacetime around a collapsing star, taking on board the Schwarzschild solution to Einstein's equations. This paper, published in the *Physical Review* in September 1939 (Volume 56, pp. 455–9), is regarded as the first modern description of the astrophysics of black holes. It was also to be the last such paper for two decades, but as Werner Israel has commented in his contribution to the book *300 Years of Gravitation*, its scope was 'breathtaking'. It used in the discussion several concepts that will become familiar friends later in the present book, using language which is exactly the same as the terminology used by relativists today. There is still no more concise, clear way of expressing our understanding of the ultimate fate of a massive star than the outline provided by Oppenheimer and Snyder in the abstract of their paper:

> When all the thermonuclear sources of energy are exhausted a sufficiently heavy star will collapse. Unless fission due to rotation, the radiation of mass, or the blowing off of mass by radiation, reduce the star's mass to the order of that of the sun, this contraction will continue indefinitely . . . the radius of the

> star approaches asymptotically its gravitational radius; light
> from the star is progressively reddened . . . The total time of
> collapse for an observer co-moving with the stellar matter is
> finite, and . . . of the order of a day; an external observer sees the
> star asymptotically shrinking to its gravitational radius.

There are three key concepts contained in those few words. The
first is that, indeed, from the point of view of an observer outside
the star, and not involved in the collapse, it does take forever for the
star to shrink within its gravitational radius (this is another name
for the Schwarzschild radius). This is what the term 'asymptotically' means in this context. The second point is the reddening of
light referred to by Hartland and Snyder. This is an effect predicted
by the general theory of relativity. Gravity, in effect, stretches the
wavelength of light that escapes from the vicinity of any massive
object. In the visible spectrum – the rainbow of colours – blue and
violet light has the shortest wavelength, and red light has the
longest wavelength. So if you start out with blue light, the
gravitational stretching will make the light more red. The process is
known as the gravitational red shift, and it only has a noticeable
effect on the waves of light coming from an object with a very
strong gravitational pull. Indeed, the effect can just be measured in
the light from Sirius B and other white dwarfs, which was one of
the clinching pieces of evidence that they really are very dense stars.

This gravitational red shift is produced in quite a different way
from the red shift in the light from distant galaxies, which is caused
by the expansion of the Universe. In the time it takes this light to
cross space to us, space itself expands, so that the light gets
stretched on its journey. This cosmological red shift is one of the
key pieces of evidence that the whole Universe is expanding, and
that therefore it was born in a big bang thousands of millions of
years ago. Because the size of the cosmological red shift in the light
from a distant galaxy is proportional to the distance to that galaxy,
it provides astronomers with a direct measure of the distances to
other galaxies. But it has nothing to do with the gravitational red
shift.

Another way of thinking of the gravitational red shift is in terms
of energy. Blue light is more energetic than red light, and the red
shift corresponds to the energy lost by the light in climbing away
from the underlying star. Although light always travels at the same
speed, it still uses up energy in escaping from the gravitational pull,

and this shows up as the red shift. For very massive, compact stars, the red shift will be so great that energy that starts out in the form of visible light will be weakened not just into red light, but beyond the visible spectrum into the form of infrared radiation, or even longer wavelength radio waves. This is what Oppenheimer and Snyder meant by saying that it is 'progressively reddened'. For radiation escaping from a collapsing star, there will come a point, as the star shrinks and the intensity of the gravitational grip at its surface increases, when *all* of the energy of the original light is used up before it can escape. The red shift has become infinite, so the light 'wave' no longer waves at all, but has faded away to nothing. Light can no longer escape from the star, which has become a black hole. This happens precisely at the time when the escape velocity from the collapsing star reaches the speed of light, as the inward falling surface of the star crosses its Schwarzschild radius – which is why the gravitational (or Schwarzschild) radius of a black hole calculated in accordance with the rules of relativity is exactly the same as the radius of a black hole calculated using the Newtonian ideas about gravity and light. But in the picture we get from general relativity, even the very last, highly redshifted, photon struggling out of the gravitational grip of the black hole still does travel at the speed of light, 300,000 km a second.

It is the final point mentioned in the abstract of the paper, though, that contains the most significant new revelation. A 'co-moving' observer is one who falls into the black hole along with the collapsing star stuff – someone sitting on the surface of the original star, if you like. And Oppenheimer and Snyder showed that, even though the collapse of the star takes forever to an observer in the outside Universe, for such a co-moving observer it is all over in a few hours. From the point of view of the star itself, the collapse into a black hole does *not* take forever. Although it was not at all clear from the paper by Oppenheimer and Snyder how these seemingly incompatible views of what was going on might be reconciled, this, as we shall see, is the key to the possibility of using black holes as shortcuts through space and time.

All that, though, was undreamed of in September 1939. Several months earlier, the problem of how stars maintain their internal fires by nuclear fusion had been solved, removing the basis for speculations about neutron cores inside stars; in the same month that the paper by Oppenheimer and Snyder appeared, Britain and

France declared war on Germany, and scientific efforts, first in Europe and then also in the US, were diverted into other channels. In 1940 Volkoff left California to work in Princeton, and Snyder took up a post at Northwestern University in Illinois. In 1942 Oppenheimer himself was given the task of choosing a site for and setting up a laboratory to carry out the research that was to lead to the development of the atomic bomb; work started at the Los Alamos laboratory the following year. None of the three pioneers (four including Serber) ever returned to research into the nature of neutron stars and black holes. Which is no real surprise, since by the time the war ended nobody except Zwicky believed that neutron stars even existed, and nobody at all believed that black holes existed. Although a few mathematicians took up the puzzle of black holes once again in the late 1950s, it was a full twenty years after the end of the Second World War that the astronomical world was startled by the revelation that neutron stars really do exist, and by the realization that if something only three times bigger than a black hole could exist, then so might black holes themselves.

Puzzling pulsars

Although the revival of interest in collapsed stars began with a chance discovery made in 1967, that discovery had its roots in a scientific development carried out during the Second World War by scientists who had been diverted from more abstract research. That development was radar. Before the war, astronomers only had observations of the Universe made at visible wavelengths, using optical telescopes. Although the fact that radio waves from space could be detected on Earth had been noticed in the 1930s (by Karl Jansky, working at the Bell Laboratories in New Jersey), there was no time for radio astronomy to develop properly before the war broke out. During the war, radar systems along the coast of the English Channel suffered from interference which was identified as radio noise coming from the Sun, and this fanned the interest of scientists involved in radar work. After the war, in many cases initially using war-surplus radar equipment, some of them began to probe the Universe at wavelengths longer than those of visible light, in the radio part of the electromagnetic spectrum. This new window on the Universe transformed astronomy in the 1950s, just as our view of the Universe was transformed repeatedly in the following decades, when, as we shall see, instruments were lifted above the atmosphere on rockets and satellites to probe the Universe at wavelengths shorter than those of visible light.

Rockets and satellites are needed to probe the Universe using short wavelength radiation – ultraviolet light, X-rays and gamma rays – because these wavelengths cannot penetrate the Earth's atmosphere. But radio waves, like light itself, can get through to the ground. And radio astronomy has one great advantage over optical astronomy. The bright blue light of the sky, which makes the stars invisible by day, is actually blue light from the Sun that has been bounced around the Earth's atmosphere ('scattered') by tiny particles in the air, so that it comes at us from all directions. Red light, with longer wavelengths, is not scattered anywhere near so much, which is why sunsets are red. This kind of scattering does not happen at radio wavelengths, so provided they are not pointed directly at the Sun radio telescopes are not dazzled in the way that our eyes, or photographic equipment attached to telescopes, are dazzled during the day (and, in any case, the Sun is nowhere near as bright at radio wavelengths as it is at visible wavelengths). So radio astronomers can observe interesting objects in the heavens twenty-four hours a day, and don't have to shut down when the Sun is above the horizon.

In fact, the Sun does influence radio waves coming to us from space. But astronomers are cunning enough to make use of this 'interference' with the signals they receive to find out more about the objects in space that emit the radio waves. There is a constant stream of material escaping from the surface of the Sun and blowing out into space and across the Solar System. This is a very tenuous cloud of gas known as the solar wind. The atoms in this wind are not electrically neutral, because even at the surface of the Sun conditions are sufficiently energetic to remove electrons from the outside of the atoms – the solar wind is, in fact, an electrically charged plasma, although very much more tenuous than the hot plasma that exists inside a star like the Sun. The density of this plasma varies, as clouds of material move out from the Sun, and one effect of this is to make radio waves passing through the plasma vary slightly in strength – they 'twinkle', or scintillate, in just the way that variations in the atmosphere of the Earth make starlight twinkle.

But stars are only affected in this way because their images are very small – just points of light. Planets, which show as tiny disks in the sky, do not twinkle, because the tiny fluctuations are averaged out over the visible disk. Of course, stars are really bigger than planets; they only look like points of light, instead of disks, because they are so far away. The same rule applies to radio sources affected by the solar wind – but it provides extra information about radio sources because, unlike stars, some of them are so large that they do show up as extended features on the sky, not just as points. Especially in the early days of radio astronomy (less so today), it was difficult to get a precise 'picture' of a radio source, a detailed map equivalent to a photograph of a star, so it was not always obvious whether the noise was coming from a point source or an extended one. Ones that twinkle, however, are definitely point sources; ones that do not twinkle are extended objects. And one inference is that twinkling radio sources must be a very long way away.

It works both ways. The fact that distant radio sources twinkle also reveals information about the nature of the solar wind, and it was this line of attack that led a young radio astronomer called Anthony Hewish to begin investigating such scintillating radio sources, as they are known, at the new radio astronomy observatory in Cambridge in the 1950s. Born in 1924, Hewish studied in Cambridge in the early 1940s and was one of a handful of war-time physicists plucked from university life to work on radar at the Telecommunications Research Establishment in Malvern, Worcestershire. In 1946 he went back to Cambridge to complete his degree, graduating in 1948 and moving straight on to research, obtaining his Ph.D. in 1952. From using scintillations as a probe of the solar wind in the 1950s, he went on to use them as a probe of the nature of radio sources, using a government grant of just £17,000 to build a new radio telescope. The pioneering radio astronomer Sir Bernard Lovell has described this award of funds as 'one of the most cost-effective in scientific history'. For it was with the new telescope that one of Hewish's research students, Jocelyn Bell, discovered the first pulsar in 1967.

Bell (now Jocelyn Burnell) had been born in Belfast in 1943, and graduated from the University of Glasgow in 1965. During the next two years, she started her Ph.D. studies in Cambridge and worked on the construction of Hewish's new telescope – which bore little resemblance to the kind of bowl-shaped antennas that the term 'radio telescope' immediately conjures up in the minds of most people. You need a special kind of telescope to observe scintillation of radio sources, because it has to be able to respond to very rapid fluctuations in the strength of the radio noise coming from space. Your eyes, for example, can see stars twinkling because they react very quickly to changes in starlight, in 'real time', to use the computer jargon; but a photographic plate, exposed for several minutes (or several hours) will show an image built up over all that time ('integrated' for all that time). A photograph will show fainter stars than you can ever see with your unaided eyes, but it will never reveal twinkling. In the same way, a radio telescope that integrates the signal from a distant object for a long time might be useful in locating the object, but it will never reveal scintillation. The new scintillation telescope designed by Hewish would operate in real time, with a very rapid response to fluctuating signals.

It was more like an orchard than the everyday image of a telescope. A field covering four and a half acres was filled with an array of 2,048 regularly spaced dipole antennas. Each dipole (a long rod aerial) was mounted horizontally on an upright, so that it was a couple of metres above the ground, making a letter 'T' with a wide cross bar. The length of the cross bar was chosen to fit the wavelength of radio noise that Hewish was interested in observing. (And, in fact, the cross bar was slightly below the top of its supporting pole; mixing the analogy, each of the dipoles, mounted across its support, looked like the crossed yard of a square-rigged sailing ship, slung across its mast.) All of these antennas had to be wired up correctly, so that any radio noise that they picked up would be combined into one signal, which was fed into a receiver where the fluctuating signals were recorded automatically in pen and ink as wiggly lines on a long strip of paper continuously unrolling from a chart recorder. By varying the way in which the inputs from each of the 2,048 antennas were added together, this system made it possible to sweep a strip of the sky running north and south, and directly overhead at Cambridge. But in order to do this, the wiring had to be just right. This tedious wiring task was obviously just the job for a research student.

The aim of the project was to identify very distant radio sources, known as quasars, by their scintillation. By the summer of 1967 (almost exactly at the time when I arrived in Cambridge to begin my own Ph.D. studies, at the then new Institute of Theoretical Astronomy), the new telescope was up and running, and showing up scintillating radio sources, as intended. You can't 'steer' a field full of antennas the way you can move a dish antenna around to look at different parts of the sky, but with a system like the one now being used by Bell for her real doctoral work you let the rotation of the Earth sweep everything around so that you cover the whole sky once every twenty-four hours. Because the scintillation is caused by the solar wind, it is strongest when the Sun is high in the sky. But the Cambridge team left their system switched on permanently – having built it, it cost very little to run, and you never know when you might find something interesting and unexpected.

On 6 August 1967, that is exactly what happened. Each sweep around the sky produced a strip of chart 30 metres long, adorned with three wiggly lines from the pen recorders. As the telescope swept around the sky, any particular source would be 'visible' to it for just three or four minutes, at a time when it was directly over-head. Bell's job was to examine kilometres of chart to find anything that looked interesting in the wiggles. When she studied the chart for 6 August, she found a tiny fluctuation, about one centimetre long, corresponding to a faint source of radio noise observed by the telescope in the middle of the night, when it was pointing in the opposite direction from the Sun. It couldn't be scintillation; most probably, it was interference from some human activity. Bell marked what she called the bit of 'scruff' on the chart, and ignored it.

But the scruff kept coming back – almost, but not quite, at the same time every night. In September, Bell had enough information to show that the scruff was always coming from the same part of the sky, reappearing at intervals not of twenty-four hours, but 23 hours and 56 minutes long. This was an important clue, since, because of the motion of the Earth in its orbit around the Sun, the apparent passage of the stars overhead does indeed repeat every 23 hours and 56 minutes, not every twenty-four hours. Just when Bell and Hewish had decided they had found something interesting and set up a high-speed recorder to monitor the fluctuations of the scruff, it faded from view for a few weeks. But in November it was back – and the new recorder showed that the scruff was actually a radio source fluctuating regularly with a period of 1.3 seconds.

This was such a surprise that, in spite of the fact that the source stayed in the same place among the fixed stars, Hewish dismissed it, once again, as interference from a human source of radio noise. Nobody had ever seen any astronomical object vary that rapidly – the most rapidly varying stars known in 1967 fluctuated with periods of about eight hours. But continuing observations gradually ruled out any possibility of human interference, and showed that the pulses themselves were extraordinarily precise, recurring every 1.33730113 seconds exactly, and each lasting for just 0.016 of a second.

Together, these measurements showed that the source of the pulses must be very small. Because light travels at a finite speed, and nothing can travel faster, fluctuations in any signals from any source can keep in step with one another only if the source is small enough to allow a light ray to travel right across it during the interval between pulses. It works like this. If a star like the Sun is so far away that we can only see it as a point of light, the brightness of the star, as we see it, depends on the brightness of different patches of the surface of the star, added together. You can imagine that the northern hemisphere of the star might get 10 per cent brighter while the southern hemisphere got 10 per cent dimmer, and the result would be that we saw no change in the total brightness of the star. We would only see the brightness fluctuate if the whole star got dimmer and brightened in step. And that can only happen if the variations happen so slowly that there is time for some sort of message to get from the north pole to the south pole, saying, in effect, 'I'm about to start getting brighter, so you'd better do so as well.' The 'message' might be a regular variation in pressure, or a repeated change in the way convection is carrying energy outward from inside the star; the point is that *whatever* the physical cause of the variation, its influence can only spread at the speed of light, or less, so the whole star can only respond in step to a disturbance if it is small enough for the appropriate message to reach every part of it before the message changes. Otherwise, some parts will be getting brighter and others dimmer, in a confused mess of variations. A precise pulse 0.016 seconds long, repeating precisely every 1.33730113 seconds, could only come from something very small indeed – about the size of a planet, or even less.

Hewish and his team had to face the very real possibility, as of November 1967, that what they had detected was indeed a signal coming from a planet – a beacon radiated by another intelligent civilization. Tongues only slightly in their cheeks, they speculated among themselves that they might have made contact with little green men, and dubbed the source 'LGM 1'. And Hewish decided to keep the lid on news of the discovery until they had carried out more observations. It was just as well that he did.

Working with another research group in Cambridge at that time, I knew, as did all the astronomers, that the radio people at the Cavendish were up to something. But just what it was they were up to, nobody could prise out of them. Well, we thought, no doubt they would tell us in their own good time. I wasn't really very interested, anyway; I was too deeply embroiled in the first real task that I had been set as a research student, developing a computer program that would describe the way in which stars oscillate, or vibrate. At the end of 1967, this seemed about as useful as spending your days wiring up a field full of antennas, and I still had no clear idea of how I might turn this work into anything useful enough to earn my Ph.D. By the end of February 1968, however, everything had changed.

Just before Christmas, Bell found another piece of scruff, coming from another part of the sky. This one turned out to be a similar source, pulsing with comparable precision to LGM 1, but with a period of 1.27379 seconds. Soon, there were two more to add to the list, with periods of 1.1880 seconds and 0.253071 seconds respectively. The more sources were discovered, the less likely the little-green-men explanation seemed. And, in any case, careful observations of the first of these objects had shown, by the beginning of 1968, no trace of the variations that you would expect if they were actually coming from a planet in orbit around a star. They must, after all, be natural. The LGM tag was quietly dropped, and Hewish decided it would be safe to go public – first with a seminar in Cambridge, to let the rest of the astronomers there in on the act, and then, almost immediately, with a paper in *Nature* (the issue dated 24 February 1968) announcing the discovery to the world.

The radio astronomers had indeed discovered a new kind of rapidly varying radio source. The title of the discovery paper was 'Observation of a Rapidly Pulsating Radio Source', and the term 'pulsating radio source' soon gave rise to the name 'pulsar', which stuck. But what *were* these pulsars that Bell had discovered?

Zwicky was right: neutron stars revealed

With the announcement of the discovery of pulsars, all hell broke loose among the theorists. A whole new kind of previously unsuspected astronomical object had been discovered, and somebody was going to make their name by finding an explanation for the phenomenon. In the discovery paper, Hewish, Bell and their colleagues pointed towards what seemed the obvious possibilities. If the radio pulses were being produced by a natural process, not by an alien civilization, they had to be coming from a compact star. Nothing else could supply the energy required to power the pulses. A star the size of a planet like the Earth had to be a white dwarf, of course; anything smaller (also allowed by the rapid pulsations) would have to be a neutron star. Many stars were known to oscillate, or vibrate, breathing in and out as a result of regular variations in the processes producing energy inside them, and varying in brightness as a result. Maybe this could also happen in compact radio stars. 'The extreme rapidity of the pulses', said the Cambridge team, 'suggests an origin in terms of the pulsation of an entire star.'* And they pointed out that the rapid speed of the fluctuation meant that the star doing the pulsating had to be either a white dwarf or a neutron star. There was one snag. Although calculations of the pulsation periods of white dwarfs had been carried out by theorists in 1966, the basic periods they came up with were no lower than 8 seconds, a little too big to explain the pulsars. On the other hand, even a simple calculation showed that neutron stars would vibrate with periods much shorter than those of the first pulsars discovered, around a few thousandths of a second. White dwarfs looked the better bet, if some way could be found to allow them to vibrate a little more rapidly than the earlier calculations had suggested.

By February 1968, my computer model of stellar pulsations was working pretty well. It would be straightforward to modify it to describe vibrating white dwarfs. What's more, the first calculations of white dwarf vibrations had used an equation of state which did not fully incorporate an allowance for the effects of general relativity. My Ph.D. supervisor, John Faulkner, pointed out that using a proper relativistic equation of state ought to allow the stars to vibrate faster. But how much faster could only be determined by carrying the calculations through the computer. Together, we adapted the computer program to the task (using, among other things, a relativistic structure equation developed by Chandrasekhar in 1964), and found that, indeed, we could get our model white dwarf stars vibrating with periods as short as one and a half

---

* *Nature*, Volume 217, p. 709, 1968.

seconds.* Our results were published in *Nature* in May 1968; when further calculations showed that by allowing for the effects of rotation, white dwarfs might vibrate as rapidly as ten times a second, it seemed for a few heady weeks that I had been involved in a major discovery. In fact, as more observations of more pulsars were made by radio astronomers around the world (a couple of dozen by the end of 1968; scores more by now), and as I pushed 'my' rotating white dwarf models to the limit, it became clear that what I had actually done was to prove that pulsars could not possibly be white dwarfs, after all.

The problem was that the fastest vibration period I could obtain, using unrealistic amounts of rotation, was still greater than the periods of some of the new pulsars being discovered. One discovery was particularly significant. It was made by astronomers using the 300-foot dish antenna at Green Bank, West Virginia – just about any kind of radio telescope can observe pulsars, once you know what to look for. They found a pulsar flicking on and off thirty times a second, near the centre of a glowing cloud of gas known as the Crab Nebula.

The high speed of the Crab pulsar, as it became known, was already enough to put the white dwarf model in trouble (and even faster pulsars have been found since). Its location, however, was even more significant than its speed.

The Crab Nebula is actually the debris from a supernova explosion – one which was observed from Earth by Chinese astronomers in AD 1054. Walter Baade, Zwicky's old colleague, had pointed out, years before, that if Zwicky was right and supernova explosions left neutron stars behind, the best place to look for a neutron star would be in the middle of the Crab Nebula. He had even identified a particular star in the Crab Nebula that he said might be the neutron star left behind by the explosion. Until 1968, almost everybody else (except Zwicky) thought he was wrong – although, as the fact that neutron stars were even mentioned by Hewish's team in the pulsar discovery paper shows, by the middle of the 1960s a few theorists were dabbling with calculations of the

---

* I was able to do this, not so much because of any outstanding ability at astrophysics but because the Institute of Astronomy had one of the best scientific computers then available, a brand-new IBM 360/44. But it is amusing to put the power of that machine in perspective. At that time, the memory of this super computer was just 128K – less than one quarter of that of the machine I am using to write this book, a Zenith SuperSport, which is itself rather elderly by the standards of 1991.

structure and behaviour of such objects. But the radio observations showed that the Crab pulsar seemed to be in the same place as the star Baade was so interested in. Further studies showed that this star was actually flicking on and off, in *visible* light, thirty times a second – something that nobody could have conceived as being possible just a few months before. A star that flickered so rapidly was beyond the wildest imaginings of the most daring theorist. Yet it did. It was indeed the pulsar, energetic enough to be detected as visible light, not just with lower-energy radio waves.

By the time those observations were made, at the Steward Observatory on Kitt Peak, Arizona, in January 1969, everyone was convinced that pulsars are indeed neutron stars. And it had also become clear that, in spite of their name, they are not pulsating, but rotating, beaming radio waves (and in some cases light) out through space from an active site on their surface. The pulses produced by a pulsar are the equivalent of a celestial lighthouse (but natural, not the product of an alien civilization), flicking its beam past the Earth repeatedly as the underlying star rotates (Figure 3.2).

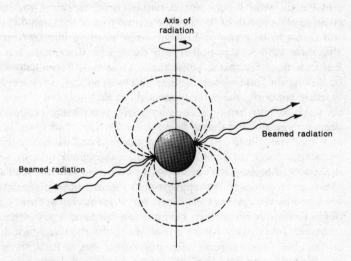

Figure 3.2 A pulsar is a fast-spinning neutron star with a strong magnetic field. Radiation is squeezed out from the magnetic poles, and as the beam rotates it flashes like the beam from a lighthouse.

There is now an overwhelming weight of evidence that this is indeed the case, and that pulsars are neutron stars spinning so fast that in many cases a spot on the equator of such a star is being whirled around at a sizeable fraction of the speed of light.

As I recall, the notion that rotating neutron stars might fit the pulsar bill was very much in the air around Cambridge in the spring of 1968, and was the subject (along with other more or less wild ideas to explain pulsars) of considerable coffee-time debate. But the person who put the idea down on paper, and published it in *Nature* in the early summer of that year (Volume 218, p. 731), was Tommy Gold, who thereby gained fame as the man who worked out the true nature of pulsars. In fact, not long before the announcement of the discovery of pulsars (and after Jocelyn Bell had first noticed the bit of scruff on her charts), Franco Pacini had published a paper in *Nature*, late in 1967,★ in which he pointed out that if an ordinary star did collapse to form a neutron star, the collapse would make it spin faster (like a spinning ice skater drawing in her arms) and strengthen the star's magnetic field, as it was squeezed, along with the matter, into a smaller volume. Such a rotating magnetic dipole, said Pacini, would pour out electromagnetic radiation, and this could explain details of the way the central part of the Crab Nebula still seems to be being pushed outwards, nearly a thousand years after those Chinese astronomers saw the supernova explode. It may seem a little unfair that to some extent Gold stole Pacini's thunder by linking the rotating neutron star idea with pulsars; however, it's worth mentioning that similar ideas about the source of energy in the Crab Nebula had been aired by Soviet researchers a couple of years previously, and that as far back as 1951 Gold had speculated, at a conference held at University College, London, that intense radio noise might be generated in the neighbourhood of collapsed, dense stars. Maybe justice was done, in a roundabout way, after all. One key feature of the application of these ideas to pulsars, predicted by Gold in his 1968 paper, was that rotating neutron stars ought to slow down slightly, spinning less fast as time passes; when measurements carried out by a team using the thousand-foot dish antenna built into a natural valley in Arecibo, Puerto Rico, showed that the pulse rate from the Crab pulsar was indeed slowing down, by about a millionth of a second per month, 'Gold's model' could no longer be doubted.

★ Volume 216, pp. 567–8.

Jocelyn Bell got her Ph.D. (and Hewish later received a Nobel Prize) chiefly for discovering pulsars; I got my Ph.D. partly for proving that pulsars could not possibly be pulsating white dwarf stars.* At the time, this seemed a distinctly unglamorous, negative thing to have done. But with hindsight, it seems much more worthwhile than I appreciated at the time. For if pulsars were *not* white dwarfs, that meant that they *must* be neutron stars, whether vibrating or rotating. I lacked the knowledge, at the time, to appreciate the significance of this and to realize that the existence of neutron stars made acceptance of the reality of black holes almost inevitable. But it is no coincidence that the term 'black hole' itself was first applied in this astronomical context in the same year that pulsars were discovered. Over the decades that followed, black holes were invoked to explain a variety of astronomical phenomena – including one that had been puzzling the theorists since 1963.

* In case you are wondering, nor can white dwarfs rotate fast enough to explain pulsars. They would be torn apart by centrifugal force long before reaching such speeds.

CHAPTER FOUR

Black Holes Abound

Watery black holes that power the most energetic objects in the Universe. An X-ray star that rings like a bell. How Hawking lost a bet. The first known black hole – and a hundred million more

When radio astronomy got going in the 1950s, astronomers already had a fairly clear idea of the nature of the Universe at large – an understanding which stemmed directly from Einstein's general theory of relativity. By describing the structure of spacetime as a coherent whole, the general theory actually provides a description of the entire Universe in terms of curved spacetime. Before the 1920s, astronomers thought that the Universe consisted of the stars we can see in the night sky, and associated material such as clouds of gas and dust in space – together making up the Milky Way system. Although individual stars might be born and die in the Milky Way, the whole system was perceived as eternal and unchanging, rather like a great forest which remains much the same for thousands of years even though individual trees within the forest live out their separate life cycles. So it came as a great surprise to Einstein, in 1917, when he applied the equations of the general theory to provide a description of the behaviour of the whole of spacetime, and found that a static, unchanging universe could not exist.

Red shifts and relativity

Such a mathematical description of the Universe (a 'universe' with a small 'u') is also known as a cosmological model; the equations do not necessarily describe 'the' Universe, but the range of possible behaviour patterns which the Universe must conform to if the general theory of relativity is a good description of reality. The equations allowed for the possibility of model universes which are always expanding, and for model universes which are always contracting, but they flatly ruled out the possibility of a static universe, seemingly in contradiction to the observations. And yet, the general theory proved triumphant in every other test that was applied to it.

The dilemma was resolved during the 1920s. Observers discovered that some of the fuzzy clouds that they could see and photograph through their telescopes are not, in fact, clouds of gas within the Milky Way, but separate star systems, comparable in size to the entire Milky Way, lying far beyond the stars we can see with our unaided eyes. They established that the Universe is much bigger than anyone had previously suspected, and that galaxies, as they are now called, many containing hundreds of billions of stars, are scattered through the void like islands scattered across the Pacific Ocean. And they also found that these galaxies are receding from one another – that the space between the galaxies is expanding, so that the Universe as a whole is expanding, exactly as required by Einstein's equations.

To my mind, this is the single most dramatic and important confirmation of the accuracy of the general theory of relativity as a description of space and time. The equations told Einstein that the Universe could not be static, and he refused (for several years) to believe the equations, suspecting that something was wrong with the theory.* But about ten years later, observations carried out quite independently, and with no expectation of testing this weird and little-known 'prediction' of the general theory, showed that the Universe is indeed expanding. The discovery came as a surprise, except to a few theorists who were aware of the implications of Einstein's work; but the equations describing what the observers had now discovered were already sitting on the pages of scientific

---

* Einstein actually introduced an extra term into the equations, called the 'cosmological constant', solely to hold his model universes still; in later life he said that this was the 'biggest blunder' of his career.

journals in the academic libraries. Ever since, relativistic cosmology, based on Einstein's equations, has been the basis for our
understanding of the Universe at large. And it is this expansion of
the Universe which tells us that long ago everything must have
been packed tightly together into a hot, dense fireball – the famous
Big Bang in which the Universe was born. The expansion of the
Universe out of the Big Bang is, in fact, the mirror image, as far as
the equations of the general theory are concerned, of the collapse of
a dense star into a black hole.

The evidence for universal expansion comes from studies of the
light of distant galaxies. When the light from a star (or any other
hot object) is spread out by a prism to form a spectrum of rainbow
colours, the spectrum usually turns out to be marked by sharp lines
at very precise wavelengths. These spectral lines come in groups,
and each group is associated with the radiation from the atoms of
one particular element. For example, hot sodium atoms (or ones
that have been 'excited' by an electric current) emit a bright yellow
light, which is familiar to us from street lighting. Studies of the
spectral lines in the light from distant stars and galaxies reveal to
astronomers what those stars and galaxies are made of. And they
also show that distant galaxies are receding from us, because the
lines in their spectra are shifted towards the red end of the
spectrum, compared with lines produced by the atoms of the same
elements here on Earth. Each set of lines (those from hydrogen
atoms, for example) forms a pattern as uniquely distinctive as a
fingerprint; in the 1920s, astronomers found that the whole pattern
is shifted bodily to the red (by a tiny amount) in the light from
distant galaxies.

The interpretation of this red shift is that the light has been
stretched on its way to us from the distant galaxy. During the time
it takes the light to reach us (which may be many millions of years),
the space between the galaxies expands, in line with the predictions
of the general theory of relativity, and the light stretches with it.
Because red light has a longer wavelength than blue light, lines that
start out with a certain wavelength end up with a longer wavelength that puts them further towards the red end of the spectrum.
This is the cosmological red shift, produced by a quite different
process from the gravitational red shift mentioned in Chapter
Three.

There are two properties of this universal expansion that are
worth mentioning in passing, even though they have no real

bearing on the story of black holes. The first is that the red shift is *not* caused by galaxies moving apart *through* space; it is the space itself that expands, carrying galaxies along for the ride, like separate raisins being moved further apart from one another in the rising dough that will make up a loaf of raisin bread. Secondly, although from our point of view here on Earth we see galaxies receding uniformly in all directions, this does *not* mean that we live at the centre of the Universe. This kind of expansion, with space stretching uniformly between the galaxies, will give exactly the same picture of symmetrical expansion for any observer, anywhere in the Universe, and there is no centre to the expansion. But the key feature of this cosmological red shift (apart from the fact that it occurs at all) is that it tells us how far away a galaxy is – the bigger the red shift the more distant the galaxy. This was the background knowledge of the Universe into which theorists tried to fit the discovery of astronomical radio sources in the 1950s.

Radio galaxies

By 1950, the radio astronomers in Cambridge had identified fifty distinct sources of radio noise coming from different parts of the sky. Unfortunately, because radio waves are much longer than light waves, it is more difficult to pin down the precise source of radio emission than it is to locate a visible star or galaxy. In effect, the image of a radio source is more blurred than the image of a star, unless you can construct a radio telescope much larger than any optical telescope. So, especially in the early days of radio astronomy, it was difficult to identify the visible counterparts to the newly discovered radio sources. Although one source in particular was identified with the Andromeda galaxy, which is another 'island' in space lying close by our Milky Way Galaxy, this was one of the fainter radio sources known, and most of the first fifty were much brighter at radio wavelengths. The natural assumption that the pioneers of radio astronomy made was that these brighter objects must be closer to us, 'radio stars' located somewhere within the Milky Way.

This raised one puzzle, in particular, although few people seem to have been bothered by it at the time. The stars of the Milky Way are concentrated in a disk, and the Solar System lies in the plane of that disk, so that the Milky Way forms a thick band of light across the sky. The newly discovered radio sources, however, seemed to be distributed at random across the entire sky. In 1951, in the same paper in which he pointed out that any radio stars must be very compact, dense stars, in order to generate so much radio noise with the aid of intense magnetic fields, Tommy Gold also pointed out that the very even distribution of these objects across the sky might well mean that they were not stars at all, but ought to be identified with other galaxies, far beyond the Milky Way. At the time, only Fred Hoyle seems to have added his weight to this idea. Most astronomers rejected it, mainly because the radio sources were so powerful – if the radio sources, each thousands of times more powerful than the radio noise from the Andromeda galaxy, were actually further away than the Andromeda galaxy, then they must be generating many thousands of times more energy in the radio part of the spectrum.

A turning point came in 1951, when Graham Smith, in Cambridge, used a technique known as interferometry to pin down the position of a very strong radio source known as Cygnus A.* With interferometry, observations made using two (or more) radio telescopes are combined to mimic the effect of a much bigger telescope. This technique has now been extended so far that it is possible to use radio telescopes on opposite sides of the world to make simultaneous observations of a source and map it as precisely as if we had a single radio telescope as big as the Earth itself. Smith's pioneering efforts were, in 1951, on a much more modest scale, but good enough to enable Walter Baade and Rudolph Minkowski, using the 200-inch telescope at the Palomar Observatory in California, to identify the source with a dumbbell-shaped object that definitely was not a star in the Milky Way. At the time, Baade thought that Cygnus A might be a pair of colliding galaxies; it is now generally accepted that it is an exploding galaxy. Either way, although it is one of the brightest radio sources in the sky, it is so faint at visible wavelengths that even the largest optical telescope on Earth only shows it as a faint blob on photographs. When Baade and Minkowski measured its red shift, they found that lines in the spectrum of this distant galaxy were shifted towards the red end of the spectrum by 5.7 per cent – a huge red shift for a galaxy, implying that it is at a distance of hundreds of millions of light

---

* The name simply means that it is the brightest radio source seen in the direction of (but actually far beyond) the constellation Cygnus.

years. The 'recession velocity'* of Cygnus A is an impressive 17,000 km per second, and it produces ten *million* times more energy at radio wavelengths than our neighbour the Andromeda galaxy does. Indeed, in absolute terms Cygnus A produces more energy at radio wavelengths than a typical bright galaxy does in the form of star light.

Once one such radio galaxy had been identified, others soon followed. Surveys of the sky carried out by the Cambridge astronomers pinpointed many more radio sources, and these are still often referred to by their catalogue numbers in these surveys – for example, the source 3C 295 is the 295th radio object listed in the third Cambridge catalogue. The 3C catalogue was completed in 1959, and contains a list of 471 radio sources; Cygnus A is also known as 3C 405. Not all of these sources are identified with galaxies. Some really are associated with objects inside our own Galaxy, including the Crab Nebula (3C 144) which (we now know) contains a pulsar. Some had still not been identified with optical objects by the early 1960s. But many of the radio sources were identified with distant galaxies, some even more remote than Cygnus A, and therefore correspondingly more energetic in order to be so radio bright to our telescopes. Where did the energy to produce all the radio noise come from? Nobody knew for sure, but in a paper published in 1961 the Soviet astrophysicist Vitalii Ginzburg made the prescient suggestion that the enormous energy required to power a source like Cygnus A might be provided by the gravitational contraction of the central part of the galaxy concerned.

There is nothing particularly mysterious about this, except for the scale of the effect Ginzburg described. If you drop a rock on to the ground, the rock gains energy as it is accelerated by the force of gravity. When it hits the ground, this energy of motion of the rock (kinetic energy) is converted into a jostling of the atoms and molecules in the rock and the ground; such a jostling produces a (tiny!) rise in temperature. Gravitational potential energy possessed by the rock in your hand has been converted first into kinetic energy and then into thermal (heat) energy. When a large cloud of gas in space collapses to form a new star, the same sort of thing happens on a grander scale. The acceleration of the individual atoms

---

* Astronomers use this term as a convenient shorthand, even though they know that the red shift is actually caused by expanding space.

and molecules in the gas cloud as they fall inward under the tug of gravity is converted into a jostling motion when the particles collide with one another, making the centre of the cloud hot. This, indeed, is how stars get hot enough inside to initiate the nuclear fusion reactions that then keep them hot for as long as the supply of nuclear fuel lasts. If you have a large enough mass collapsing in this way, then, as Ginzburg pointed out, you can generate just about as much energy as you want to. The stronger the gravitational pull of the object on to which the gas is falling, the easier it is to release energy. And within a couple of years of Ginzburg putting forward the proposal that radio galaxies might be powered in this way, astronomers began to realize that in some cases they might be dealing with very strong gravitational fields indeed. It all began with the discovery of what seemed, at first, to be genuine radio stars but turned out to be a previously unsuspected kind of astronomical phenomenon that includes the most distant objects visible from Earth, some with recession velocities in excess of 90 per cent of the speed of light, seen by light that left them more than ten billion years ago, five billion years before the Sun and the Earth had even formed.

Quasars

The first step towards the discovery of the objects now known as quasars was made in 1960, with the identification of the optical counterpart to another of the sources in the third Cambridge catalogue, 3C 48. First, astronomers using the world's biggest steerable dish antenna, the famous radio telescope at Jodrell Bank, linked into an interferometer system, found that the radio noise from this source was coming from a tiny point on the sky, less than four seconds of arc across (this is about the angular size of Mars when it is farthest from Earth). Then, armed with this information, Thomas Matthews, of the California Institute of Technology, used an interferometer telescope in Owens Valley to pin down the location of the source as accurately as possible. His colleague Allan Sandage then used the 200-inch telescope to take a long-exposure (actually 90 minutes) photograph of that part of the sky. The photograph showed what seemed to be a blue star, even smaller than the limit set by the Jodrell Bank observations, precisely at the position of the radio source.

The first thing astronomers do when they are investigating a visible object is to take its spectrum. This they duly did for 3C 48, and found that although the spectrum was richly marked by lines, the patterns the lines made were unlike those seen in any other star. In particular, the observers could find no trace of lines corresponding to hydrogen, even though hydrogen is by far the most common element in all stars.

Sandage announced the discovery in December 1960, at the annual meeting of the American Astronomical Society. But he and his colleagues were so puzzled by the spectrum of 3C 48 that they didn't even publish their findings in the proceedings of that meeting. Apart from a brief mention in the report on the meeting carried by the magazine *Sky and Telescope* a few weeks later, nothing about the discovery was published until 1963 – and even then, Matthews and Sandage didn't know what it was they had found. The *Sky and Telescope* report had commented that:

> Since the distance of 3C 48 is unknown, there is a remote
> possibility that it may be a very distant galaxy of stars; but there
> is general agreement among the astronomers concerned that it is
> a relatively nearby star with most peculiar properties.*

That was still the consensus at the beginning of 1963. Within a few months, however, investigations of yet another of the 3C sources had shown that the consensus was wrong.

These investigations stemmed from a new trick to determine the positions of radio sources, worked out by the British astronomer Cyril Hazard. As the Moon moves across the sky, it passes in front of a few stars, and any other objects, that happen to lie within the band on the sky traced out by the Moon. Such an event – rather like an eclipse – is known as an occultation. Hazard pointed out, in 1961, that if a radio source is occulted in this way, then by carefully timing the moment when the radio source goes 'off the air' and the moment when it reappears, it would be possible to locate the position of the source from the known position of the Moon on the sky. All you have to do is draw two curves on your star map, one marking the leading edge of the Moon at the time when the source disappears, the other marking the trailing edge at the time the source reappears. The two curves will cross each other twice, but at least this pins down the location of the source to one of two points on the sky, and usually the radio measurements are good enough to distinguish between them.

In fact, you can do even better than that, if you are lucky. Hazard realized that during 1962 there would actually be three lunar occultations (in April, August and October) of a radio source known as 3C 273, which had not been identified with any visible object. With three occultations, the technique should give an unambiguous, pinpoint position for the source. Hazard and colleagues in Australia used a then new radio telescope at the Parkes Observatory to monitor these occultations, and once again a photograph taken with the 200-inch telescope showed that there was what seemed to be a blue star, this time with a jet of material apparently being ejected from it, at the position of the radio source.

This 'star' also had an unusual spectrum, crossed with an unfamiliar pattern of lines. But Maarten Schmidt, a Dutch-born astronomer working in California, who obtained the first spectrum of 3C 273, found an explanation for the strangeness. He realized that one particular set of four lines could be explained as a characteristic 'fingerprint' of hydrogen – but with a red shift of just under 16 per cent. Accepting this value of the red shift, other lines in the spectrum also fell into place. Jesse Greenstein, who had obtained the spectrum of 3C 48 in 1961 and worked alongside Schmidt in California, immediately took another look at his old data, and found that the strange spectrum of that object could also be explained by a large red shift – an even more staggering 37 per cent, corresponding to a velocity of recession of 110,000 km a second and a distance of several *billion* light years.

The 'discovery' paper for 3C 273 from the Parkes team, a paper by Schmidt announcing its red shift, and a paper by Greenstein and Matthews announcing the red shift of 3C 48 all appeared in the same issue of *Nature* in 1963.* Schmidt pointed out in his paper that there are only two ways in which such a huge red shift could be produced, either by the gravitational stretching of light or by the expansion of the Universe. But 'it would be extremely difficult, if not impossible,' he said, to account for the particular observed spectrum in the case of 3C 273 by the gravitational red shift, and 'at the present time . . . the explanation in terms of an extragalactic origin seems most direct and least objectionable.' Thirty years on, it still does. There is now an overwhelming weight of confirmatory

* Volume 197, pp. 1037, 1040 and 1041.

evidence that these objects which look like stars but have huge red shifts, hundreds of which have now been identified, really are at the cosmological distances those red shifts imply.

Cosmic powerhouses

Although they were named 'quasistellar' objects in the years immediately following their discovery, this soon got shortened to quasars. We now know, as well as we know anything in astronomy, that a quasar is the bright core of a very distant galaxy, producing an enormous output of energy, a hundred times brighter (or more) than ordinary galaxies like Andromeda, to make it visible across billions of light years of space. Yet rapid variations in the output of quasars show, using the same reasoning that limits the size of pulsars, that the energy is coming from a region only about as big across as our Solar System. They are cosmic powerhouses without equal. But how can such a small source generate so much energy? Ginzburg had provided a clue, and on page 533 of the same volume of *Nature* with the 3C 273 papers in 1963 there was a paper by Fred Hoyle and Willy Fowler which suggested that the kind of energy required could be provided only by the release of gravitational energy as an object with a mass of a hundred million suns collapsed to 'the relativity limit' – in other words, to its Schwarzschild radius. The Schwarzschild radius for an object with this much mass would, indeed, be about the same as the radius of the Solar System. But it took astronomers another ten years to accept that the cosmic powerhouses in quasars really are supermassive black holes.

This was partly because, in a sense, quasars were discovered too soon – before even the identification of pulsars, which showed that neutron stars must exist, and that therefore black holes almost certainly existed as well. During the 1960s, various more or less bizarre theories to account for the way quasars generate energy were discussed, and almost all were eliminated. But one suggestion, which had been put forward in some detail as early as 1964 (by the Soviet researchers Yakov Zel'dovich and Igor Novikov, and by Ed Salpeter in the United States), stood the test of time. Although this model has been refined since, especially by Donald Lynden-Bell, Martin Rees and their colleagues in Cambridge, the essence is still the same. It gives us a picture of a central black hole, as big across as the Solar System and with a mass of a hundred million

suns, lying at the heart of a young galaxy* surrounded by a
swirling disk of material from which it gradually swallows up
matter. Each mouthful of matter that it swallows produces a release
of gravitational energy, heating the surrounding material. And
because it is surrounded by a disk of matter, energy from the region
just outside the black hole will be squirted out along its poles, often
producing jets like the one seen in 3C 273 (Figure 4.1). This is,

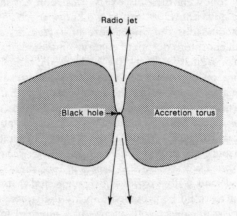

Figure 4.1 *The accretion disk of material swirling around a
black hole only leaves two narrow channels, at the 'poles' of the
hole, through which matter and energy can escape.*

indeed, a lot like the way that pulsars radiate energy; but nobody
knew that when quasars were discovered.

There are two remarkable features of this kind of black hole,
from the point of view of the present book. The first is that in spite
of its huge mass, such an object has a density roughly the same as
that of our Sun – less than twice the density of water. It is almost
exactly the kind of 'dark star' envisaged by Michell in the eighteenth century! The second astonishing discovery is that the conver-

* The galaxy has to be young, because if we see a quasar by light which has taken several
billion years on its journey to us, it left not long after the Universe emerged from the Big
Bang. It is a reasonable guess that when the Universe was young galaxies were full of gas
which had not yet formed into stars, and that this is what 'fed' the black hole. That also
explains why older galaxies, which are closer to us, are not active in this way – even if they
harbour a supermassive black hole, there is no spare gas around to feed the beast.

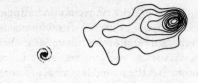

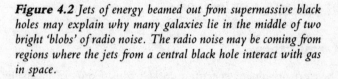

Figure 4.2 Jets of energy beamed out from supermassive black
holes may explain why many galaxies lie in the middle of two
bright 'blobs' of radio noise. The radio noise may be coming from
regions where the jets from a central black hole interact with gas
in space.

sion of gravitational energy into radiation as the black hole
swallows up matter is so efficient that, even though a quasar may
shine a hundred times more brightly than an ordinary galaxy, it
only needs to consume an amount of matter equivalent to two or
three times the mass of our Sun each year in order to maintain that
output. Since galaxies contain hundreds of billions of stars, it is
quite easy to see how a quasar can continue to shine brightly for
millions, or even hundreds of millions, of years.

What is even more important, though, is that there is absolutely
no alternative way to account for the energy output of quasars
except with the aid of these cosmic powerhouse gravity machines.
Their existence is very strong evidence indeed that black holes
really do exist – so much so that it is now widely accepted by
astronomers that many galaxies, including our own Milky Way,
may have a black hole at the core. There is evidence of all kinds of
energetic activity in the hearts of galaxies, from quiet ones like our
own up to quasars, and all levels of activity in between. The
difference between our Galaxy and a quasar may simply be that the
black hole at the heart of the Milky Way has a mass of 'only' a
million suns, and that it has gobbled up all the surrounding
material, so that it has no disk of gas left to feed on.

All of this, though, became established only in the late 1970s –
some of it only in the 1980s. Why did it take so long, even after the
discovery of pulsars? Partly because there were new developments
in the theoretical understanding of black holes during the 1960s and
1970s (more of this in Chapter Five); but chiefly because it was only

in the early 1970s that astronomers found unambiguous evidence for the existence of black holes with masses only a few times that of our Sun, among the stars of our Milky Way Galaxy itself. It was the proof that such 'stellar mass' black holes exist that persuaded the doubters that the supermassive black hole theory of quasars was on the right lines. But the foundations of *that* discovery were actually laid as far back as June 1962, when Hazard and his colleagues were still in the throes of trying to pin down the location of 3C 273 using the lunar occultation technique.

X-ray stars

The electromagnetic spectrum doesn't consist only of visible light and radio waves. It includes infrared radiation, ultraviolet light, X-rays and gamma rays – all waves that obey Maxwell's equations and travel at the speed of light, but none of which (unlike radio waves and visible light) can penetrate the atmosphere of the Earth. To see what the Universe looked like at these wavelengths, astronomers had to hoist instruments above the obscuring layers of the atmosphere, first with balloons and rockets, then with orbiting satellites. As early as 1948, simple measuring devices flown by American scientists on captured German V2 rockets, left over from the Second World War, had shown that the Sun is a weak source of X-rays, as well as radio waves and light. X-rays are a more energetic form of radiation than visible light, with shorter wavelengths, and would be produced in profusion only by an object much hotter than our Sun. Astronomers were not surprised that the Sun itself should produce some X-rays, especially at times when there are flaring bursts of activity on the solar surface. But if other stars produced a similar amount of weak X-radiation, there would be no hope of detecting stellar X-rays even from the top of the Earth's atmosphere. Although the solar X-ray activity was studied intermittently throughout the 1950s, nobody suspected that it might be possible to detect X-rays from further away in space than our Sun. Indeed, the step that opened up the science of X-ray astronomy came when researchers actually set out to look for X-rays from a celestial object that is *closer* to us than the Sun – the Moon.

Of course, the Moon is far too cold to produce its own X-rays. But some scientists reasoned that energetic particles emitted by the Sun (the so-called solar wind) must strike the surface of the Moon, and they speculated that the impact of these particles might energize atoms in the lunar material, making them emit X-rays at character-istic wavelengths. If this happened, it would provide a means of finding out what material the Moon is made of, by a kind of X-ray spectroscopy. An experiment intended to identify X-rays from the Moon was launched from White Sands, in New Mexico, on board an Aerobee rocket on Monday, 18 June 1962. It was a failure, in the sense that it failed to detect any X-rays from the Moon;* but it was a spectacular and unexpected success in that it found a bright source of X-rays, seemingly coming from a point on the sky.

During the flight, in which the instrument package was above the atmosphere for less than six minutes, the spinning detectors also found a faint background of X-rays coming from all directions in the sky, and hints of at least one other faint individual X-ray source. But it was the bright source that was the sensation. It seemed to be an object far outside the Solar System, and this was confirmed by later rocket flights, which showed that it was always in the same part of the sky, in the direction of the constellation Scorpius. It was soon dubbed Sco X-1 (meaning the first X-ray source seen in the direction of Scorpius). Something out there was so hot and ener-getic that it was producing X-rays in profusion, easily visible across interstellar space – a true X-ray star.

Other X-ray sources were soon discovered. But at first, nobody knew what kind of star might be producing the X-rays. X-ray astronomy suffered from a similar problem, especially in the early days, to radio astronomy – the detectors could not pin down precisely the locations of the objects they were studying. In the case of radio waves, this is because they are dealing with very long wavelengths. X-rays have very short wavelengths, so in principle accurate X-ray telescopes need not be as large as accurate radio telescopes, or even optical telescopes like the 200-inch. But the instruments carried on the first rocket flights were tiny, and could hardly be called telescopes at all – they had about as much directional sensitivity as a wide-angle camera lens. What's more, they were moving. On the flight that discovered Sco X-1, as well as soaring up out of the atmosphere and back down again in the span

---

* Indeed, the predicted lunar X-rays were at last discovered only towards the end of 1990, more than twenty-eight years after that pioneering Aerobee flight, by instruments on board the orbiting X-ray satellite ROSAT.

of a few minutes, the rocket was spinning twice a second, ensuring that the detectors scanned right round the sky but making it even harder to decide precisely where the X-rays they detected were coming from. So X-ray astronomers borrowed one of the tricks of radio astronomy to identify at least one of their sources. The first to be identified with a known visible object was in the Crab Nebula.

It was a rocket flight in April 1963 that pinned down the rough location of Sco X-1, and the same flight also showed a much fainter source of X-rays coming from the general direction of the Crab Nebula. The obvious guess was that the source lay precisely in the Crab Nebula, which was, after all, the site of a supernova explosion. Herbert Friedman, of the US Naval Research Laboratory, suggested that the X-rays might be coming from a neutron star left behind by the supernova, and that Sco X-1 might also be a supernova remnant. This marked the beginning of the revival of the idea that neutron stars are made in supernova explosions, put forward thirty years earlier by Zwicky and Baade, and soon to receive a dramatic boost with the discovery of pulsars.

Friedman was lucky. At the time his team discovered the X-ray source that they suspected might be linked with the Crab Nebula, Hazard's lunar occultation technique had just had its spectacular success in the identification of 3C 273. What's more, the Crab Nebula itself lies in the right part of the sky to be occulted by the Moon. This happens only once every nine years – but the next such occultation was due on 7 July 1964. There was just a nice amount of time for Friedman's group to plan an Aerobee flight to monitor the X-rays from the source during the occultation.

It wasn't quite as easy as I make it sound. The launch had to be perfectly timed so that the five minutes or so of viewing time coincided with the occultation – a delay in the countdown would wreck the experiment. And the Aerobee launcher itself was far from perfect – six launches in a row, up to the critical date, had failed because of defects in the control system. But the one that mattered worked perfectly, showing that the X-rays are indeed coming from a point at the centre of the nebula, a point which Friedman confidently identified as a neutron star, though this conclusion was not fully accepted until the discovery of pulsars. Even if pulsars had never been discovered, however, continuing studies of X-ray stars would soon have shown that they must be associated with very dense objects.

Celestial powerhouses

The reason for Friedman's confidence that at least some X-ray sources must be neutron stars concerns the amount of energy produced by these stars. As I have mentioned, dropping matter into a strong gravitational field is a very efficient way to release energy. The falling matter accelerates to very high speeds, and when it hits the surface of the object it is falling on to, this kinetic energy is converted into heat. Dropping matter on to the surface of a neutron star would be a very good way to make the surface of the star hot – so hot that it would, indeed, radiate energetically at X-ray wavelengths. This was clear from general principles (as long as you believed in the existence of neutron stars) even in 1964. But what evidence was there to back up the idea?

The evidence came following the identification of Sco X-1 with a visible star. As X-ray detectors were improved, the location of the source was pinned down more precisely, until in March 1966 it had been located accurately enough for the optical astronomers to get in on the act. In June that year (still a full year before the first hint of the existence of pulsars), Japanese observers identified a peculiar star close to the suspected position of Sco X-1. Studies with the 200-inch telescope soon showed that this star flickered in an extraordinary manner, changing in brightness from one minute to the next. Astronomers keep photographic records of the sky as a matter of routine, and the star also showed up in these photographs, dating back to the 1890s, which indicated that it also varies on longer timescales. But the most dramatic feature of the identification of the star associated with Sco X-1 was that it is about a thousand times brighter in the X-ray part of the spectrum than it is in visible light – and radiates about 100,000 times as much energy as the *total* output from the Sun.

The flickering, flaring, and overall energy output of Sco X-1 can all be explained in one package. It requires that there are *two* stars associated with the X-ray source – that it is a binary system, with two stars orbiting each other, locked in a mutual gravitational embrace. In this situation, if one of the stars is very dense and compact, while the other is larger, with a more diffuse atmosphere, gas from the atmosphere of the large star will be torn away from it by tidal effects, and attracted to the small star. As the gas spirals down on to the small star, it will form a swirling disk of material, generating heat within the disk itself as gravitational energy is converted into kinetic energy – a quasar in miniature, in many ways, although even in 1969 the equivalent quasar model had not yet become fully established in the minds of all astronomers. The small, dense star is surrounded by a very hot gas, or plasma, which radiates at X-ray wavelengths (as well as a relatively modest amount of visible light) and is constantly being renewed by gas falling in from the larger star.

In 1969 I was able to explain the flickering in the light from Sco X-1 in terms of vibrations in the hot plasma surrounding the underlying star in just such a binary system. These flickers sometimes show short bursts of regular, periodic fluctuations, before becoming more messy; the bursts of regular variation usually follow large flares, which can be interpreted as a result of an extra large glob of material being dumped on to the X-ray star by its companion, releasing an extra burst of energy and also setting the plasma cloud vibrating, rather like a bell that has been struck by a giant hammer.

The oscillation of such a hot plasma depends on its physical properties (such as temperature and density) and on the strength of the gravitational field in which it is held. Studies of the spectrum of Sco X-1 showed what the plasma was like, so the oscillation period of the plasma showed the strength of the gravitational field. The numbers I came up with in 1969 were no surprise – the flickering of Sco X-1 showed that the underlying star might be a white dwarf, could not possibly be an ordinary star like our Sun, and is most probably a neutron star. Nobody was particularly amazed or impressed by my conclusions, because they came two years after the discovery of pulsars, when astronomers no longer doubted the existence of neutron stars. But it is interesting that such studies of X-ray sources provide completely independent evidence that neutron stars exist, and that neutron stars might therefore have been discovered in this way if Sco X-1 had been identified a little sooner, or if pulsars had been identified a little later. And the binary model of X-ray sources turned out to be a key ingredient in the next great step forward, which occurred in the 1970s and provided the best ever proof that one specific source is indeed a black hole.

The prime candidate

X-ray astronomy stepped into space properly on 12 December 1970, with the launch of a satellite dedicated to making X-ray observations of the sky. Instead of a rocket flight lasting a few minutes, the detectors on this satellite were in orbit around the Earth, able to scan the skies continuously for as long as they kept working, and as long as the satellite stayed in orbit. And, after eight years of development, the detectors were far more accurate and sensitive than anything flown on the early Aerobee rocket launches.

The first X-ray astronomy satellite was launched from a platform in the sea just off the coast of Kenya, in East Africa. The site was chosen because it is just south of the Equator, and by launching a rocket from the Equator into a west-to-east orbit it is possible to take advantage of the rotation of the Earth to give it a boost into space, through a kind of slingshot effect giving a push start worth about a thousand miles an hour. That mattered because the rocket, provided by NASA but launched by an Italian team, was a relatively small one, called a Scout. The exact launch date was chosen because it was the seventh anniversary of the independence of Kenya from British colonial rule, and to mark the occasion the satellite was named Uhuru, which is the Swahili word for freedom. The choice of name was doubly appropriate, since the launch freed astronomers, for the first time, from the struggle of trying to observe the heavens through a blanket of atmosphere.

The impact of Uhuru and the observations it made during its three-year active life can only be compared with imagining the impact on science if the Earth had been eternally shrouded in cloud until 12 December 1970, and then the clouds had suddenly rolled back to reveal the stars. Uhuru found that the sky is, indeed, covered with X-ray sources, some identifiable with visible stars, others not. Some of these sources were clearly, like Sco X-1 and the Crab source, part of our Milky Way Galaxy; others, equally clearly, were associated with distant galaxies. The Universe was a far more violent and energetic place than astronomers had imagined, even after the discovery of Sco X-1, and continued monitoring of many of these sources, by Uhuru and its successor satellites, showed them to be just as variable as that archetypal X-ray source.

The story of X-ray astronomy since 1970 would fill several books – and has. I shall concentrate on just one of the sources probed by Uhuru and its successors, the source which was actually also seen, more faintly than Sco X-1, on the pioneering Aerobee flight in 1962, and which, although very variable in strength, is sometimes the second-brightest object in the X-ray sky, after Sco X-1 itself. It lies in the direction of the constellation Cygnus, and, as the first X-ray source found in that part of the sky, it is known as Cygnus X-1.

Some of the X-ray stars monitored by Uhuru showed regular variations just like X-ray versions of pulsars. They are explained in terms of the rotation of a neutron star which is energized by infalling matter from a companion star, just like the accepted model for Sco X-1. The reason why we do not see such regular pulsations from Sco X-1 is, presumably, because the Solar System doesn't happen to lie in the region swept out by the 'lighthouse beam' from that particular source (and, by the same reasoning, for every radio pulsar we see there must be several, perhaps many, that we cannot detect because their radio beams do not happen to pass across the Solar System). But Cygnus X-1 was not one of these X-ray pulsars. Nor did it look exactly like Sco X-1, although there were similarities. Like Sco X-1, it showed rapid variations in X-ray brightness, with occasional large flare-ups, and with short intervals in which there would be a more or less regular, rapid flickering with periods ranging from a few tenths of a second to a few seconds. These short-lived periods of regular flickering are much faster than those in Sco X-1, however, which means that the plasma that is vibrating is in the grip of a stronger gravitational field. Since Sco X-1 is almost certainly a neutron star, this immediately begins to look interesting – and the rapidity of the flickering also tells us directly, from the usual light speed argument, that the source of the X-rays must be less than 300 km across.

With the improved estimate of the position of Cygnus X-1 provided by Uhuru, astronomers began to look for an optical counterpart to the source. Unfortunately, there were many stars in that part of the sky, and no obvious way to pick out the one they were interested in. Maybe, though, the radio astronomers could help. At that stage in the development of X-ray astronomy, the detectors being used on the satellite were less accurate, as far as pinning down the positions of the sources were concerned, than radio telescopes on the ground, which could make use of the ever-improving technique of interferometry. In June 1970 and again in March 1971 radio astronomers at the Green Bank Observatory, in West Virginia, pointed their instruments in the general direction of Cygnus X-1, as part of a general survey of the sources studied by Uhuru. They found nothing; but on 13 May 1971 they did detect radio noise coming from that part of the sky. Meanwhile, radio astronomers at the Westerbork Observatory, in the Netherlands, had also discovered that a radio source had suddenly 'turned on' in the right part of the sky, some time between 28 February 1971 (when they looked and found nothing) and 28 April 1971 (when they looked again and found the source). Together, the two sets of observations showed that the radio source had turned on some time between 22 March and 28 April. The Uhuru data showed that just about at the time the radio source appeared, the X-ray output from Cygnus X-1 dropped to a quarter of its previous value. Nobody knew why this had happened (and nobody has yet come up with an entirely satisfactory explanation), but all that mattered was the coincidence in timing. Clearly, the 'new' radio source had to be associated with Cygnus X-1, where energy had somehow been diverted from X-radiation into radio noise. And the radio astronomers provided two estimates of the position of the source, both of which coincided almost exactly with the location of an ordinary-looking star that had been catalogued many years before at Harvard College Observatory, and was known as HDE 226868.

HDE 226868 is an ordinary B-type star, larger and brighter than the Sun – a blue supergiant. It looks so faint to us, however, that it must be thousands of light years away. Optical astronomers immediately turned their telescopes on this star. Louise Webster and Paul Murdin, working at the Royal Greenwich Observatory in England, and Tom Bolton, of the David Dunlap Observatory in Canada, soon found, independently of each other, that HDE 226868 is, in fact, a member of a binary system. The blue supergiant star is orbiting, once every 5.6 days, around an invisible companion. Now, a blue supergiant cannot possibly have a mass much less than 12 times the mass of the Sun, and most stars like this have masses 20 or 30 times that of the Sun. Using Newton's and Kepler's laws, it is simple to calculate the mass of the companion object if HDE 226868 has a mass of 12 solar masses and an orbital period of 5.6 days – the companion must then have a mass of three times that of the Sun. If HDE 226868 is any more massive, then its companion must be correspondingly more massive, in order to hold it in such a tight orbit (the distance between HDE 226868 and Cygnus X-1 is only one fifth of the distance between the Earth and the Sun). In other words, the companion object, the source of the X-rays, must be at least three times as massive as the Sun. It cannot be another bright star, or we would see it – and, remember, the flickering shows that all the mass is packed into a sphere no more than 300 km across. Its mass exceeds the Oppenheimer-Volkoff limit, and, as both teams of observers had realized by 1972, that makes it a black hole.

Since then, the evidence that Cygnus X-1 is a black hole has got stronger. The details of the improved understanding involve spectroscopic studies of the system, and analysis of the orbital motion. The bottom line, as of 1987, was drawn by black hole theorist Roger Blandford, in his contribution to the volume *300 Years of Gravitation*. Summing up the evidence, he concluded that the *minimum* possible mass of the blue supergiant is 16 solar masses, implying that the X-ray source is a black hole with a mass seven times that of the Sun, while the *most probable* mass of HDE 226868 is 33 solar masses, implying a mass for Cygnus X-1 of 20 solar masses. 'The case that Cyg X-1 has a mass in excess of the Oppenheimer-Volkoff limit is', says Blandford, 'pretty strong, and, importantly, has strengthened significantly since 1972.' It is very hard to prove anything about an object thousands of light years away from us, but he concludes that, rather than quibble about the very faint possibility that there is something fundamentally wrong with the interpretation of the observations (if so, it means we understand so little about how stars work that we might as well give up astronomy altogether), 'it is surely more productive at this stage to accept the evidence and proceed.' Accept the evidence, that is, that black holes really do exist. And although Cygnus X-1 remains the prime candidate, the inference is that there are hundreds of millions of these objects in our Galaxy alone – even though very few of them have actually been detected.

A profusion of possibilities

As of 1991, there is literally a handful of X-ray sources – just five – for which there is good evidence that the gravitational pull of a black hole provides the energy which makes the plasma we see hot enough to produce X-rays. In at least two of these cases, the evidence is nearly as good as in the case of Cygnus X-1 itself. Twenty years on from the launch of Uhuru, it is a little disappointing that there are no more firm candidates than this. After all, since the discovery of the first pulsar in 1967 astronomers have identified roughly five hundred neutron stars. But that comparison is a little misleading. Out of those five hundred pulsars, only a handful are known to be in binary systems. An isolated pulsar, spinning on its

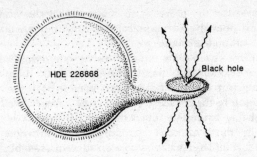

Figure 4.3 *On a much smaller scale than Figure 4.2, the first black hole to be definitely identified explains the origin of X-rays from close to the star known as HDE 226868. Matter torn off from the star by the black hole forms a swirling disk in which gravitational energy is turned into heat and produces X-rays. The X-ray source is known as Cygnus X-1.*

own in space, can still be detected by the radiation beamed outward by the strong magnetic fields at the surface of the neutron star. But an isolated black hole, with no infalling matter to feed off, lives up to its name. It is black, and undetectable. Indeed, the fact that we see roughly the same number of good black hole candidates as we do binary pulsars suggests that there may be as many isolated black holes around in our Galaxy as there are isolated neutron stars.

How many is that? The five-hundred-odd known pulsars represent just the tip of the iceberg, according to current astronomical thinking. A pulsar, after all, does not live forever. The ones we see are relatively young and active neutron stars, and as they age they slow down and radiate less energy, eventually fading away into invisibility. Astronomers have a good idea of the way stars evolve, and how many of them turn into supernovas and explode every thousand years or so in a galaxy like our own. The Galaxy contains a hundred billion stars, and it has been around for thousands of millions of years. Even though only a few of those stars explode as supernovas every millennium, that still means that there may be as many as four hundred million 'dead' pulsars around in the Galaxy, and a fairly cautious estimate made by Roger Blandford suggests that there may be a third as many – about a hundred million – isolated black holes scattered across the Milky Way. If so, the chances are that the nearest one is about 15 light years away – almost next door, by astronomical standards, but tantalizingly out of reach, and undetectable.

Since there is nothing unusual about our own Galaxy, this calculation also suggests that every galaxy in the Universe must contain stellar mass black holes in comparable profusion. And this is in addition to the evidence that all large galaxies like the Milky Way probably harbour a much more massive black hole at their heart. The most cautious astronomers would argue that the evidence that *all* large galaxies contain supermassive black holes is still only circumstantial. But the launch of ROSAT in 1990 provided astronomers, for the first time, with pictures of the sky at X-ray frequencies as detailed as photographs in visible light taken through optical telescopes.* Within a few months of its launch, the satellite had found 24 X-ray quasars (that is, quasars visible in ordinary light that also produce X-rays) in a patch of sky covering just one third of a square degree, corresponding to 72 such sources in each full square degree. It is impossible to explain the production of X-rays by quasars in quantities prodigious enough to be detected at the distances implied by quasar red shifts without invoking the black hole energy generation mechanism. On the other hand, it is easy to produce that much energy with the aid of a black hole with a mass a few hundred million times that of our Sun. Even allowing for the likelihood that the particular patch of the sky studied in this early survey by ROSAT contains an unusually large number of such objects (the region was, indeed, chosen for investigation because studies by earlier X-ray satellites had suggested something interesting might lie in that part of the sky), the inference is that there are many thousand giant black holes around in the Universe, potentially visible to us (when ROSAT gets around to photographing them all) as X-ray quasars. Almost all quasars are, on this evidence, active in the X-ray part of the spectrum.

Even the black holes involved in these energetic processes may not, though, be the last word as far as mass is concerned. During the 1980s, astronomers were intrigued by the discovery of several pairs of identical quasars; each pair actually seems to be a double

* Among other things, these precision observations confirm (not that confirmation was really needed) the identification of Cygnus X-1 with HDE 226868. The name 'ROSAT', incidentally, is derived from that of Wilhelm Röntgen, who discovered X-rays in 1895. It has taken less than a hundred years to move on from that discovery to a satellite capable of taking X-ray photographs of the heavens.

image of a single object. The explanation for this imaging is that light from a very remote, single quasar is being bent by gravity around some intervening massive object, so that it arrives at Earth from two slightly different directions, producing two images. This is known as the gravitational lens effect; it is the same process, on the grand scale, as the light bending that provided the first test of the general theory of relativity, in the eclipse of 1919, and it was predicted by Einstein himself (Figure 4.4). In some cases, the

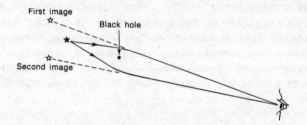

Figure 4.4 *A black hole can act as a gravitational lens, producing multiple images of a background star or galaxy by the light-bending effect.*

gravitational lensing may be due to the presence of a large galaxy lying along the line of sight between us and the distant quasar. But in at least three cases there is no sign of a bright galaxy in the right place for its distortion of space to produce the lens effect. It is possible, though by no means yet proven, that in these cases the lensing is being caused by the presence of individual super-supermassive black holes, each weighing in at around a thousand billion solar masses.

All in all, most – virtually all – astrophysicists now regard black holes as a natural feature of our Universe. They seem to be a natural product of the evolution of massive stars, as well as playing a key role in the behaviour and evolution of quasars and galaxies. As Blandford puts it, 'astronomers and physicists have got used to the idea.' They have got used to it because of an overwhelming weight of observational evidence coming from studies of pulsars, binary X-ray sources, quasars, and galaxies with energetically active centres. Thirty years ago, before anything was known about any of these phenomena, no astrophysicists took the notion of black holes seriously. Going hand in hand with the observational progress made since the early 1960s, the mathematicians have more than kept pace with developments, refining and improving their theories of black holes to explain each new phenomenon, and to suggest, as we shall see, possibilities far more exotic than anything the observers have yet discovered. Everything I have described so far is now conventional wisdom, the image of black holes that astrophysicists have got used to. What the relativists still have up their sleeve, however, is as scary and unwelcome to many astrophysicists today as the notion of black holes itself was thirty years ago. As the observers would be the first to acknowledge, even in the dark days before the identification of the first known quasar, just a few relativists were at work, taking the occult art of the theory of black holes beyond the Oppenheimer-Volkoff limit. Soon, those investigations were to take them, in theory at least, to the edge of time itself.

CHAPTER FIVE

Darkness at the Edge of Time

D_ark-age theorists. How a mathematical hobbyist provided a new perspective on black holes and the existence of the Universe. The year that black holes got their name, and the inevitability of the singularity. Why black holes have no hair, but turn astronauts into spaghetti. How Hawking (with a little help) put the heat into black holes. The edge of time unveiled_

During the dark ages of black hole research, between 1939 and 1963, only a few theorists kept the flag flying. After the work by Oppenheimer and Snyder was published, just at the outbreak of the Second World War, no further progress was made with understanding the equation of state of dense matter until 1957. By then, however, not only had physicists developed a better understanding of the forces at work inside the nucleus, they also had available an unprecedented new tool, in the form of electronic computers. The improved physics and the new computer technology were combined by researchers at Princeton University in 1957 to calculate the behaviour of very dense stars in more detail than ever before. The leader of the Princeton team that carried out the work was John Wheeler, a physicist who had been born in 1911, and already had an impressive track record in physics. Among other things, he had worked with Niels Bohr, the quantum pioneer, in Copenhagen in the 1930s, and in the 1940s he was the research supervisor and collaborator of Richard Feynman, now widely regarded as the greatest theoretical physicist of the past fifty years.

In his early black hole investigations, Wheeler's assistants at Princeton were Kent Harrison (who helped out with the physics) and Masami Wakano (who handled the computing side). Together, they brought the work by Chandrasekhar on white dwarfs and the work by Oppenheimer and Volkoff on neutron stars into one unified framework, and confirmed that there was no way to stabilize a cold star with more than a certain mass.*

At the time, however, Wheeler's interpretation of this result was much the same as Eddington's response to Chandrasekhar, a quarter of a century earlier. He assumed that stars must somehow lose mass, in order to avoid being left with more than the critical amount at the end of their lives. In June 1958, when he reported this work in Brussels to a major scientific meeting, known as a Solvay conference, Wheeler said that:

> No escape is apparent except to assume that the nucleons at the centre of a highly compressed mass must necessarily dissolve away into radiation – electromagnetic, gravitational, or neutrinos, or some combination of the three – at such a rate or in such numbers as to keep the total number of nucleons from exceeding a certain critical number.

'Nucleon' is simply the generic term which describes both protons and neutrons, so the comment, as phrased, would apply equally to white dwarfs and to neutron stars. Oppenheimer, who was in the audience, disagreed with Wheeler's conclusion, and asked:

> Would not the simplest assumption about the fate of a star more than the critical mass be this, that it undergoes continued gravitational contraction and cuts itself off from the rest of the universe?†

But Wheeler was not convinced, and continued to believe for some time that the extreme physical conditions in objects as dense as neutron stars would allow some loophole which prevented continued gravitational collapse. After 1958, the notion of neutron stars slowly began to gain respectability; but in the western world, at least, black holes were taken seriously by physicists only after the discovery of quasars, when astrophysicists began to appreciate that

---

* This was the beginning of the work which revised the mass limit found by Oppenheimer and Volkoff, and which led to the modern estimate for the limit of about three solar masses.
† Both quotations from the Brussels meeting are taken from Werner Israel's contribution to *300 Years of Gravitation*.

a supermassive black hole would form not at superdensities but out of matter with only about the same density as water, where no exotic processes might be imagined acting to evaporate away the excess mass before the black hole could form.

Interestingly, though, even in the early 1950s collapsed objects were accepted as a standard, text-book phenomenon in the Soviet Union. The work by Oppenheimer and Snyder was taken at face value from the outset, and by the 1960s a whole generation of students had been brought up on the idea. This is one reason why, once quasars and then pulsars were discovered, so much of the physical insight that led to an explanation of the new phenomena came initially from Soviet researchers, such as Zel'dovich. Before quasars were discovered, however, there was one last development on the mathematical front which was to have repercussions in black hole research that echo down to the present day.

New maps of space and time

This new development resolved a puzzle which had confused physicists ever since Schwarzschild came up with his solution to Einstein's equations in 1916. What is the physical significance of the Schwarzschild horizon around a black hole? At first sight, it seemed to many researchers that this must be a real, physical barrier – an edge to space. After all, if an object fell towards the Schwarzschild surface, time would run slower and slower for the falling object as it got closer and closer to the horizon, until time itself stood still at the horizon. It would take forever for a falling object to reach the horizon, so, obviously, nothing could cross the horizon.

Or look at it another way. The escape velocity from the horizon surface is the speed of light. This means, turning the equations around, that any object falling on to the horizon from a great distance will be travelling at the speed of light when it gets there, and gravity will still be pulling it inwards, trying to accelerate it further. Since nothing can travel faster than light, it seems that falling objects must, somehow, just pile up at the horizon, and never penetrate it. Just as there is a singularity at the centre of the black hole – a point of infinite density – so it seemed that there must be a real, physical singularity at the Schwarzschild horizon, as well.

But all of these arguments are based on the point of view of an observer who sits outside the hole, watching objects fall on to its surface. From the point of view of an observer who is falling into the hole, nothing unusual at all happens at the horizon! The equations tell us that, according to falling clocks, it takes only a short time to fall right through the horizon, and penetrate the interior of the black hole. Only when the falling astronauts tried to get back out into the Universe would they discover that they had been trapped by the black hole's gravity, and were destined to plunge into the singularity at its centre. By the 1930s, relativists had realized that the Schwarzschild surface does not represent a physical singularity after all; it looks like a singularity in Schwarzschild's solution to Einstein's equations because of the way he chose his metric. The singularity is an artefact of the coordinate system used to measure spacetime around the black hole, much as the way we measure latitude on Earth seems to produce singularities at the North Pole and the South Pole, even though there are no physical singularities there.

If you set out on a journey travelling north, for example, you will eventually come to the North Pole itself. There, it is impossible to travel any further north. Because of the way we define our coordinate system, all directions from the North Pole are south. But this does not mean that there is an edge to the planet at the North Pole, nor does it mean that polar expeditions must pile up at the Pole with nowhere left to go. You can even continue your journey in the same direction that you have been going – you pass right across the North Pole, and find that you are now heading south, even though you have not turned around.

Something very similar happens at the Schwarzschild horizon around a black hole. You can pass right through the horizon, and keep going in the same direction. But something strange does happen at the horizon, even though this will not be immediately apparent to you. Although it looks as if you are simply travelling on in the same direction, towards the central singularity, the roles of space and time have been interchanged. Outside the hole, we have freedom (within certain limits) to move about as we wish to in space, but we are inexorably carried through time from the past to the future at a rate of 60 seconds every minute. Within the hole, a traveller would have freedom, within certain limits, to move about *in time*, but would move inexorably through space to hit the central singularity. More of this later; for now, I want to bring the understanding of black hole geometry up to date.

It is the *mathematics* that breaks down at the Schwarzschild surface, not the physics, so all that the relativists needed was a better mathematical description of what is going on. But that was easier said than done – especially since, on closer inspection, it turned out that what Schwarzschild had found was not one solution to Einstein's equations, but a pair of solutions, rather like the positive and negative 'roots' of a simple quadratic equation. The equations that describe the ultimate collapse of an object into a black hole can be reversed, and then describe the expansion of an object *out* of a singularity (sometimes referred to as a 'white hole'). These solutions are equivalent to the cosmological solutions Einstein found, which describe the Universe at large – the discovery that the Universe must either be expanding or contracting, but cannot stay still. Indeed, the expansion of the Universe out of the big bang is exactly the process described by the 'other' set of black hole equations.

A way to understand all this in physical terms, together with a coordinate system which made it easier to see what was going on, was developed initially in the 1950s, and in its final form in the 1960s. The first steps were taken by Martin Kruskal, one of Wheeler's colleagues at Princeton, in the mid-1950s. Kruskal was a specialist in plasma physics, but along with some of his colleagues he formed a group that taught themselves general relativity, more or less as a hobby. Kruskal found a coordinate system in which the structure of a black hole could be described in one smooth set of equations, joining the flat spacetime far outside the hole on to the highly curved spacetime inside without even a mathematical hint of a singularity at the Schwarzschild horizon (in essence, this coordinate system describes things from the perspective of a light ray plunging into the black hole). But when he showed his calculations to Wheeler, a couple of years before Wheeler carried out his study of dense stars with Harrison and Wakano, he expressed no interest in it, and Kruskal never bothered to publish it. By 1958, Wheeler had realized the importance of Kruskal's mathematical discovery, and he began to spread the news of it at scientific meetings. But Kruskal, engrossed in his own research, had lost interest, and still didn't publish the result formally. In the end, Wheeler himself wrote up the work, put Kruskal's name on the paper, and sent it off to the *Physical Review*, where it was published in 1960. Later still, Roger Penrose, of Oxford University, improved on Kruskal's representation of the structure of space and time associated with a black hole. For mathematicians, it is the Kruskal metric that is the key to understanding black holes; for physicists, the key insight comes from a pictorial representation, known as the Penrose diagram.

This diagrammatic representation actually kicks off from Minkowski's insight which led to the description of flat spacetime in terms of four-dimensional geometry. Because we can't draw in four dimensions, and because all three dimensions of space behave in the same way as each other, while time is the odd dimension out, relativists often represent events that are occurring in spacetime in terms of lines drawn on a two-dimensional diagram, like a graph, in which time is measured up the page and one dimension representing space is measured across the page. Such a simple spacetime (or Minkowski) diagram is shown in Figure 5.1. This is a slightly more sophisticated version of the kind of diagram represented by Figure 2.4; by choosing each unit of time up the page as one year, and each unit of length across the page as one light year,

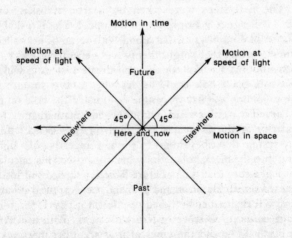

Figure 5.1 *A more sophisticated version of the spacetime diagram (Figure 2.4, p. 46) relates the location of events in spacetime to the speed of light. From 'here and now', you can travel to anywhere in the future, and you can get information from anywhere in the past. But you can never know anything about, or visit, the regions marked 'elsewhere'.*

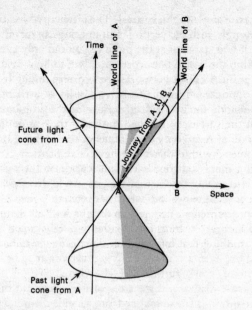

Figure 5.2 *In fact, it is better to think of the regions of space and time accessible from 'here and now' in terms of the future and past light cones. From the point A, an observer can know nothing at all about events at point B. But as time passes, there comes a moment when information about B enters the future light cone of the observer at a point on the world line 'above' A. At that moment, the observer can receive a signal from B.*

we can make sure that the path of a light ray through spacetime is represented on such a diagram by a line at 45° to the vertical.

The key feature of such a diagram is that it provides a pictorial image of how any observer interacts with the Universe at large. Each point on the diagram represents a moment in time, *and* a position in space. If any observer sits still at any point in space, they will always be the same distance from the time axis, and as time passes all they do is age. A line representing the (rather boring) history of such an observer (the observer's 'world line') just goes vertically up the page. If the observer moves about, though, the world line will be wiggly. If you move from A to B in Figure 5.2, and take a year to make the journey, your world line will be tilted at an appropriate angle to the vertical. The faster anything travels, the bigger the angle with the vertical. But nothing can travel faster than light. So, if you start out at the point A you can only ever travel to points within the world lines corresponding to light rays travelling out from point A. The two world lines corresponding to light rays going in opposite directions form two sides of a triangle; if you imagine spinning the whole diagram around the time axis, they will sweep out the surface of a cone, known as the 'future light cone'. The observer at A can only ever influence things that go on in parts of the Universe within the future light cone of A.

Similarly, there is a cone extending back into the past from the observer at A. Only things that happen inside that past light cone can have any influence on what happens at point A – which, remember, represents a moment in time as well as a point in space. Although there are regions outside the two light cones, there is no way that anything that happens out there can ever influence, or be influenced by, events at point A. Wherever you may be, spacetime is divided into 'past', 'future' and 'elsewhere'.

In a Penrose diagram, distant regions of space and time (all the way out to infinity) are all mapped into a single diamond shape on a Minkowski-type diagram, so that they can all be fitted into the page. This is a fairly straightforward mathematical ploy, no more difficult than using Mercator's projection to map the spherical surface of the Earth on to a rectangular piece of flat paper. Although it does (like Mercator's projection) distort the picture somewhat (so that, for example, the region of spacetime inside a black hole gets half as much space on the diagram as the entire Universe outside), what matters is that it shows how different regions of spacetime are connected to one another, and which regions can or cannot be visited from a chosen point without violating the speed-of-light limit.

The first stage in representing spacetime in the presence of a black hole in this way looks like Figure 5.3. The entire Universe outside the hole is represented by the diamond. The inside of the black hole is represented by the triangular shape at the top right, while the singularity itself is represented by a jagged line which covers all of space within the hole at a certain time – it marks the edge of time itself. Within the hole, all roads lead to the singularity. The boundary of the hole, the event horizon, is marked by an arrow-head, showing that it can only be crossed in one direction. And it is easy to see that in order to get out of the hole, you would have to

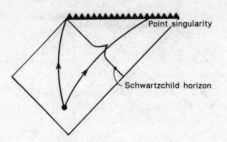

Figure 5.3 *A spacetime map of the entire Universe can be represented by a diamond shape, rather like the way a representation of the surface of the Earth can be mapped on to a square sheet of paper. As usual, the future is 'up the page' and the past is 'down the page', and motion at the speed of light would be at an angle of 45°. In this representation, a black hole is denoted by the triangular region of spacetime 'alongside' the Universe. The point singularity is denoted as a horizontal line, showing that anything which falls into the hole across the Schwarzschild horizon must hit the singularity as it travels inexorably into the future. The arrowhead on the line representing the Schwarzschild horizon indicates that it can only be crossed in one direction – going into the black hole.*

follow a world line tilted at more than 45° to the vertical, travelling faster than light, which is impossible.

But this is only half the story. Where is the white hole solution to the equations? It comes, *complete with an entire extra universe*, in the full version of the Penrose diagram for a Schwarzschild black hole, shown in Figure 5.4. Now, there is a white hole in the past, from which things can emerge into either universe, but into which nothing from either universe can ever fall. And both universes share the black hole singularity in the future. There is no way for travellers to pass from one universe to the other, although suicidal astronauts diving into the black hole from each universe could meet, and briefly compare notes, before they were annihilated in the singularity. If the Universe, born in a big bang, is destined one day to collapse into a black hole singularity (and there are compelling reasons to think this is the case, discussed in my book *In Search of the Big Bang*), then this Penrose diagram is the best pictorial representation of the entire life cycle of our Universe. Which

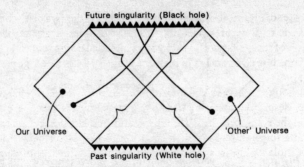

Figure 5.4 *In fact, the full representation of the way a black hole is connected to the rest of spacetime requires an extra universe on the 'other side' of the hole, and a singularity in the past, known as a white hole. But there is no way to get from our Universe to the extra universe without travelling faster than light or backwards in time.*

means, among other things, that the existence of a second universe must be taken seriously. This might not be much of a worry if there were no way to communicate with it, even in principle, as the simple Penrose diagram suggests. But this diagram represents the structure of spacetime for a simple, non-rotating black hole – the kind described by Schwarzschild's solution to Einstein's equations, and therefore known as a Schwarzschild black hole. Since real black holes are likely to be rotating (and stellar mass black holes might very well rotate faster than pulsars), maybe this version of the Penrose diagram is just a little *too* simple. In Chapter Six, I'll be looking at what happens *inside* the event horizon when we add in the effects of rotation; but for the rest of this chapter I want to concentrate on the way in which realistic black holes, both rotating and non-rotating, are likely to interact with the Universe at large.

Black holes in a spin

The explosion of interest in black holes after 1963 (including Penrose's development of the diagram that now bears his name) stemmed from two things – the discovery of quasars, and the discovery of a solution to Einstein's equations that describes the nature of a rotating black hole. Before going on to look at the

Penrose diagram for such an object, and the way it connects different regions of spacetime, it's worth looking at the physical nature of these beasts – each of which harbours not one event horizon, but two, and in which the singularity is not a point but a ring.

Although Schwarzschild found his solution to Einstein's equations, describing a static black hole, just about as soon as Einstein found the equations, in 1916, it took another 47 years before anyone found a solution to Einstein's equations which described a rotating black hole. This is a measure of just how much complexity is contained within the equations of the general theory of relativity, and just how difficult they are to solve; mathematicians do not imagine that they have yet plumbed all of the depths of the equations, by any means, and more solutions (contributing more surprises) may yet emerge from them. The particular difficulty with the puzzle about rotating black holes, which was eventually solved by a New Zealander, Roy Kerr, working at the University of Texas, is that a rotating mass drags spacetime around with it as it rotates. The effect had been known about, as a theoretical projection of Einstein's equations, for a long time; but until Kerr's work in 1963 nobody knew what the consequence of this dragging of spacetime in the vicinity of a black hole would be. It took a further twelve years, incidentally, to prove that the Kerr solution to Einstein's equations is the only one that describes rotating (and electrically neutral) black holes, just as it has been proved (by Werner Israel, in 1967) that the Schwarzschild solution is the only one that describes non-rotating, uncharged black holes. The Schwarzschild solution is, in fact, a special case of the Kerr solution, with the rotation set at zero.

If you fell into a spinning black hole at one of its poles, you wouldn't notice the way in which the hole drags spacetime around with it. You would only reach the Schwarzschild horizon at the usual distance from the singularity (a distance depending only on the mass of the hole) and pass through it on a one-way journey to extinction. Round by the equator, though, the dragging effect would be very apparent. As well as the inward tug of gravity pulling you towards the hole, there would be a sideways drag, carrying you around with the rotation of the hole. Within a certain distance from the horizon, it is impossible to stand still, no matter how powerful the motors on your spaceship may be. Although you can still use your rockets to stop you falling into the hole itself, and can even escape back into the Universe outside, you will be dragged sideways, no matter how hard the rockets blast. The limiting distance within which this inevitable dragging takes place is known as the static limit, and it is further out from the event horizon the closer you get to the equator, as you move round from the pole. The static limit marks a boundary known as the static surface, surrounding the rotating black hole like a fat doughnut (Figure 5.5),

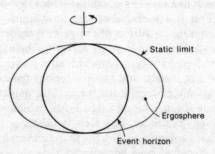

Static limit

Ergosphere

Event horizon

Figure 5.5 *Spacetime near a rotating black hole is dragged around by the rotation. The region that is affected is called the ergosphere, because Roger Penrose showed how energy could be extracted from it. The outer boundary of the ergosphere is called the static limit; the inner boundary is at the event horizon (or Schwarzschild surface) of the black hole.*

and the region between the static surface and the event horizon is known as the ergosphere (from a Greek word meaning 'work'), because of a curious property of black holes discovered by Roger Penrose.

Penrose has been one of the key players in the black hole game since the early 1960s, and we shall come across his work again. He was born in 1931, and obtained his Ph.D. from Cambridge University in 1957, spending the next nine years researching and teaching in London, Cambridge, Princeton, Syracuse and Texas before settling down at Birkbeck College in London in 1966. In 1973 he moved to Oxford, where he is still Rouse Ball Professor of Mathematics, with interests (some of them related in his best-selling book *The Emperor's New Mind*) that extend far beyond the black hole investigations for which he is best known. His special insight into the nature of rotating black holes came in 1969, when he showed how such an object could be used as a source of energy.

By that time, incidentally, researchers such as Penrose were actually using the term 'black hole'. The name was first applied to collapsed stars by John Wheeler in 1967, when he tried it out on his colleagues in an informal way before giving it its first full public airing at the New York meeting of the American Association for the Advancement of Science on 29 December. This astrophysical use of the expression black hole first appeared in print in the January 1968 issue of the magazine *American Scientist*, and immediately caught on, replacing earlier terms such as 'frozen star' and 'collapsar'. The timing, in the wake of the discoveries of quasars, pulsars and X-ray stars, was perfect. As Wheeler says in his book *A Journey into Gravity and Spacetime*, 'the advent of the term *black hole* in 1967 was terminologically trivial but psychologically powerful. After the name was introduced, more and more astronomers and astrophysicists came to appreciate that black holes might not be a figment of the imagination but astronomical objects worth spending time and money to seek.'

Penrose, though, was one of the few researchers already fascinated by black holes, and in need of no further encouragement to maintain his interest. He says that the idea of a way to extract energy from a rotating black hole came to him on the train, on his way to meet up with his students in London, while trying to think of something new to tell them about these objects. In the Penrose process, as it has become known, an object falls into the ergosphere, where it breaks into two separate pieces. One piece heads across the event horizon, but travels in the opposite direction to the rotation of the hole (it is only possible to travel against the spin provided the chunk of matter is heading *into* the hole). The other piece heads out of the ergosphere, travelling in the direction of the rotation – but it travels out much faster than the whole of the original object came in, having received a boost in energy from the dragging of spacetime around the hole (having been 'worked on' by the ergosphere). It is as if the outgoing chunk of the original object has received a 'kick' from the chunk that falls into the hole, reminiscent of the way a rifle kicks against the shoulder of a marksman when it is fired. But the kick is more powerful than the recoil of any rifle in the outside Universe. If the trajectory of the infalling object is chosen carefully, and the timing of the split is just right, the piece that emerges from the ergosphere will actually carry more energy out than the whole object took in across the static surface. The energy has come from the rotation of the black hole, which has actually been slowed by a tiny amount as it has been forced to swallow up an object travelling counter to its spin. Indeed, the *mass* of the black hole decreases by a tiny amount, as some of its mass energy is converted into energy of motion, giving the outward-moving chunk of the object a boost through the rotating region of spacetime that is the ergosphere.

This can be explained in terms of the energy of motion possessed by a particle in the ergosphere. For a particle at some distance above the Earth, or above the Sun, the energy of motion is zero when the particle is stationary, hovering in one place (perhaps with the aid of a rocket blasting to oppose the force of gravity). This is simple commonsense. But be wary of applying commonsense where black holes are concerned. In the ergosphere, because of the dragging of spacetime a particle has to orbit slowly around the black hole, in the opposite direction to the way the hole is rotating, in order to have zero energy of motion. And if it moves against the spin of the hole faster than this critical velocity, it doesn't gain energy, but *loses* it – it actually has *negative* energy of motion. It is the addition of this negative energy (equivalent to the subtraction of positive energy) to the hole that makes it lose mass; since the total mass-energy of the system must stay the same, the lost energy is matched by the increase in energy of motion of the outgoing chunk of the original object.

But however much mass is added to the black hole, it turns out that a particular combination of the total mass of the hole and its angular momentum (a measure of its spin) always stays the same, or increases – it can never decrease. This property was the most important discovery made by a research student at Princeton, Demetrios Christodoulou, who followed up Penrose's discovery of the black hole energy process in 1970. The irreducible quantity is known as the 'irreducible mass' of the black hole, and the square of the irreducible mass is proportional to the surface area of the event horizon. In other words, the area of the event horizon can only stay the same or get bigger. This turned out to be a key insight, developed in the 1970s, as we shall see, by Stephen Hawking and his colleagues.

The Penrose process is not really a very practical way for people to obtain energy, even if we did know where to find a rotating black hole (although it only takes a little imagination to see how this kind of process, operating naturally, might help to account for the enormous outbursts of energy from quasars). I cannot, however, resist mentioning another bizarre, but impractical, prediction of the equations – that it would be possible to use a rotating black hole to amplify light, turning it into a kind of black hole bomb. In the early 1970s, several physicists pointed out that a similar effect to the Penrose process would increase the energy of a beam of light passing through the ergosphere, in a process known as superradiant scattering. If you imagine surrounding a rotating black hole with a spherical mirror that has a tiny hole in it, then shining a weak beam of light in through the hole, the light could pass through the ergosphere, getting amplified as it did so, bounce off the mirror, and pass through the ergosphere again and again, getting more powerful all the time. If you left the hole in the mirror open, radiant energy would build up inside (at the expense of the rotation of the black hole) until it came streaming out of the hole in the mirror in an intense beam. But if you sealed up the hole in the mirror, radiant energy would carry on building up until the mirror itself exploded outwards – the black hole bomb.

Intriguing though these speculations are, however, what goes on inside the event horizon of a black hole is even more intriguing. In particular, what is the meaning of the singularity? Does such a bizarre object inevitably form, even if it is hidden inside an event horizon and cannot be seen? And is there any way in which a singularity might exist *without* hiding behind the cloak of an event horizon, so that it could interact with the Universe at large? Even before he turned his attention to ways of extracting energy from the ergosphere of a rotating black hole, Penrose had been probing these mysteries, starting out with the question of whether singularities are an inevitable requirement of the general theory of relativity.

Singularities rule

If you think of black holes forming from collapsed stars at the end of their lives, with a density of matter greater than the density of matter in the nucleus of an atom, the idea of a singularity forming at the heart of the black hole may not seem too great a leap for the imagination, even if Eddington always found the idea abhorrent, and Wheeler took several years to come round to it. If we are already dealing, in theory, with conditions more extreme than anything ever encountered on Earth, it is no real surprise to find that the equations predict strange and extreme phenomena. But when we are dealing with a black hole as big as the Solar System, containing the mass of a hundred million Suns, but with a density scarcely greater than that of water, the idea of a singularity at the heart of the hole becomes more questionable. Can a large sphere made out of water – no matter how much water – *really* imply the existence of a singularity somewhere in its heart? If you had a great glob of water floating around in space, not quite massive enough to make a black hole, and then added the extra pint or two to take it over the limit, would a singularity inevitably form inside the hole, just because of the extra couple of pints you had added?

It sounds ridiculous – but remember that although the *average* density of a black hole might be the same as that of water, that doesn't mean that it really is a sphere made of water. A mass of a hundred million Suns in a volume no bigger across than our Solar System would quickly collapse under its own weight, no matter what it is made of.* What the equations tell us is that a black hole consists of an event horizon, a singularity, and *nothing at all* in between the two. From the outside, you can measure the mass of the hole from its gravitational attraction, and the speed with which it rotates. If it has an electric charge, you could measure that, as well. But those three properties are *all* you can ever measure. There is no way to tell what the matter that went into the hole was before it was swallowed up behind the horizon – whether it was a star, a great glob of water, or a pile of frozen TV dinners. There is no way to distinguish a black hole made of stellar material from one made of anything else, a property summed up by relativists in the expression 'black holes have no hair', coined by Wheeler and his colleague Kip Thorne in the early 1970s.

But that still allows for a lot of difference between compact, superdense black holes and supermassive, low-density black holes – at least as far as any outside observer is concerned. The most obvious example of this concerns the fate of the intrepid astronaut

---

* Of course, such a sphere made of water would collapse and fragment into individual stars, which would get hot inside from the release of gravitational energy and burn for as long as their nuclear fuel (the hydrogen from the water) lasted. The ultimate collapse would be postponed, and would take place only after nuclear burning was finished; but this is a detail that can be ignored in order to keep the story simple.

who ventures near, or even across, the event horizon. So far, I've talked glibly of somebody making such a journey, and what they might see, without mentioning the inconvenient fact that they are likely to be torn to bits by the gravitational and tidal forces they encounter. An observer who is falling freely, feet first, towards a black hole does not feel any weight, of course; but because the observer's feet are closer to the hole than his or her head, the feet feel a stronger tug of gravity and accelerate faster. As a result, the observer's body is stretched. At the same time, because everything attracted by the hole is being squeezed towards a point at the centre, the observer's body is squashed sideways. This simultaneous stretching and squeezing is exactly the same as the tidal forces that move water about on the surface of the Earth as the Moon and Sun exert their gravitational influences – but in the case of a black hole with a mass a few times greater than that of our Sun, the tidal forces are very extreme. For a black hole with ten solar masses (and therefore with a Schwarzschild radius of just under 30 km), the tidal forces acting on the unfortunate astronaut would be ten times larger than the force of gravity at the surface of the Earth when the astronaut was still 3,000 km away from the hole, and probably still unable to see it against the background of stars. Even at that distance, it would feel like hanging from a trapeze with ten other people hanging from your ankles – *and* being squeezed sideways at the same time. Long before the doomed falling observer even reached the event horizon, he or she would be 'spaghettified', and no longer in a fit state to notice what was going on.

But supermassive black holes are different. Choose one with a big enough mass, and correspondingly large radius, and the tidal forces you would experience as you fell in through the event horizon would be no worse than the forces your body experiences when taking off in an aeroplane. The intrepid astronaut really could, in these circumstances, survive to study the interior of the black hole. It would be largely a waste of time, though, since in a matter of minutes the astronaut would plunge on to the central singularity, and the same spaghettification process would take place, but inside the event horizon instead of outside. At least, the same spaghettification will take place if there really is a singularity inside the hole. Can we be sure that there is?

In fact, we can be sure. Roger Penrose proved this, and published the proof as long ago as 1965, by calculating the way in which gravity inside a black hole must distort the light cones for any points of spacetime in the hole. For a supermassive black hole made out of a uniform lump of matter, the same in all directions (spherically symmetric) and collapsing under its own weight, it is pretty obvious that the situation is just like that of a collapsing star, only more extreme, and that a singularity must form. The fact that the event horizon forms at a larger radius, where the tidal forces are less extreme, is really only a minor detail. But Penrose wanted to check whether a singularity must form if the cloud of material that formed the supermassive black hole was not spherically symmetric. Suppose the hole were made of, literally, a hundred million stars like the Sun, falling together in some messy and complicated way. Might it be possible for the particles of which the stars were made to somehow dive into the centre of the cloud and pass by each other without colliding, then move outward again from the centre of attraction, like a comet swinging in past the Sun and back out into space? Then, the density inside the hole might get very large, without ever actually becoming infinite.

It seemed a plausible idea, but a mathematical investigation of the behaviour of light cones inside the hole ruled it out. The kind of light cone I have described so far, with straight sides, corresponds to flat space. But, as we know from Einstein's work, gravity bends space so that light rays follow curved geodesics. Light rays that start out from any point inside the black hole will begin to diverge, but the light-bending effect of gravity will act like a lens, bending the rays back towards each other. If the light rays are in a strong enough gravitational field, then they will be bent so much that they converge back upon themselves, meeting up at a point. This will happen for any light rays emitted from any point inside the horizon of a black hole – it has to, or some of the light could escape from the hole. And Penrose showed that if this is the case then it is an absolute requirement of general relativity that there must be a singularity somewhere inside the event horizon. The singularity does not necessarily have to be exactly the same kind of singularity that you would get from the smooth, symmetrical collapse of a spherical star, but, as Penrose put it in a radio broadcast in 1973, 'tidal effects which approach infinity will occur, producing a region of spacetime where infinitely strong gravitational forces literally squeeze matter and photons out of existence.'*

The idea was taken up, in 1965, by a research student in

* Published in *Cosmology Now*, edited by Laurie John, BBC Publications, London, 1973.

Cambridge, Stephen Hawking. Penrose had proved that any object undergoing gravitational collapse must form a singularity. Hawking realized that by turning the equations around it might be possible to prove that the *expanding* Universe must have been born *out of* a singularity. He spent several years refining the mathematics, in collaboration with Penrose, and in 1970 they published a joint paper which proved that the Universe we observe must indeed have been born in a big bang singularity if general relativity is correct. With this important discovery behind him, in the early 1970s it was Hawking, more than anyone else, who was associated with dramatic new developments in the theoretical understanding of black holes which went hand in hand with the dramatic new observations of objects such as Cygnus X-1. His most famous discovery is that black holes explode – a discovery that calls into question a hypothesis which most physicists would dearly like to be true, but for which there is actually no evidence at all.

Defeating the cosmic censor

The fact that Penrose had proved that behind every event horizon there lies a singularity was not too disturbing, even if a singularity is, by definition, a place where the laws of physics break down and anything can happen. If we can never see the singularity, it doesn't really matter. But what would matter would be if we found a singularity that was not cloaked by the respectable screen of an event horizon. Such a naked singularity would not just be an extreme gravitational influence, sucking objects into its grip. It is a feature of the way in which the laws of physics break down at a singularity that it could defy gravity itself, spewing energy and matter out into the Universe. It would, in fact, be more like a white hole than a black hole. What's worse, this outpouring could take any form, as Hawking and others established in the 1970s. Just as there is no way to tell if a black hole is made out of star stuff or frozen TV dinners, so a naked singularity doesn't care if the matter it spews out is in the form of star stuff or frozen TV dinners. It is much more likely that it would be star stuff – fundamental particles like protons and neutrons.* But what comes out of a naked singularity is produced entirely at random, so that there is a small,

---

* And, in case you are wondering, there have been suggestions that this could explain the outpouring of matter and energy we see in quasars, although this hypothesis is not very fashionable among the theorists at present.

but real, probability that such an object might suddenly eject a replica of the Taj Mahal, all those TV dinners that I have already mentioned, or several copies of the book you are now holding, printed in green ink on red paper.

Physicists are unhappy at this prospect. Having proved that there is no such thing as an empty event horizon, in the sense that each one contains a singularity, Penrose speculated that there may be no such thing as a naked singularity, in the sense that each one is concealed by an event horizon. It seemed neat and logical, and has become known as the Cosmic Censorship Hypothesis – the idea that nature abhors a naked singularity. Unfortunately, nobody has ever been able to prove that the infallible cosmic censor is actually at work in the Universe. Clifford Will summed the situation up in his contribution to the volume *The New Physics* (edited by Paul Davies): 'There is no convincing proof of the Cosmic Censorship Hypothesis. There is not even general agreement on how to formulate the vague notion of censorship into terms that can be translated into mathematics.' Indeed, since we know that the Universe itself emerged out of a singularity in the big bang, such evidence as there is suggests that the Cosmic Censorship Hypothesis is wrong. In the 1990s, more evidence to this effect came from computer simulations of the way in which non-spherical objects actually collapse.

A singularity, in this context, is most simply understood as a place where density and gravity become infinite. It doesn't have to be a mathematical point, but could be a line, or even a sheet, of infinite density. If any kind of singularity were able to interact with the outside world, physics as we know it would break down.

Kip Thorne, of CalTech, suggested in 1972 that black holes with horizons could only form, in general, if an arbitrary mass got sufficiently compacted in all directions at once. Thorne is one of the handful of black hole specialists around; born in 1940, he obtained his Bachelor's degree from CalTech in 1962, and his Ph.D. from Princeton in 1965, arriving on the scene just at the time of the revival of interest in collapsed objects. He has been a professor at CalTech since 1970, and has worked particularly closely with Wheeler. The proposal he made shortly after he took up that post was equivalent to saying that, whatever its actual shape, a collapsed object would only form a black hole if it could pass through a hoop with the appropriate critical radius, whatever the orientation of the object; this is known as the 'hoop conjecture'. In 1990, Stuart

Shapiro and Saul Teukolsky of Cornell University in Ithaca, New York, carried out numerical simulations of gravitational collapse using the Cornell supercomputer. Their calculations suggest that Thorne was right and that cosmic censorship can be violated.

Shapiro and Teukolsky calculated the effects of the collapse of slightly non-spherical objects, or spheroids, some of which start out slightly prolate (cigar-shaped), others slightly oblate (flattened spheres, like the Earth). Compact spheroids do indeed collapse to form black holes, getting small enough in all directions to pass through a hoop with the appropriate Schwarzschild radius. But this is not the case when the spheroids are initially large.

Large prolate objects collapse to a spindle, with a linear singularity extending like a spike out through the poles of the collapsed object. Oblate spheroids collapse initially to a pancake, but pass right through this state, first becoming prolate and then collapsing to a spindle. In both cases, the linear singularity extends far beyond the boundaries of the appropriate hoop – so there is no concealing event horizon shutting it off from the rest of the Universe.

The calculations take full account of general relativity, and suggest that spindle singularities without event horizons can form in the Universe. Although gravitational radiation carries away some of the mass during the collapse, the total mass energy lost is much less than 1 per cent, so this cannot save a massive object from disappearing into a singularity. According to the Cornell team, the gravitational potential, gravitational force, tidal force and kinetic and potential energies all blow up in the hearts of these objects, even though they remain open to view. It seems that you can have a singularity without an event horizon, even though you cannot have an event horizon without a singularity. Small wonder that Will says that 'one of the most important unsolved issues in classical general relativity is the validity (and even the meaning) of the Cosmic Censorship Hypothesis.'

Of course, the study by Shapiro and Teukolsky is 'only' a computer simulation, and it may be that the researchers have missed something in their calculations. Perhaps all collapsing objects do conceal the nakedness of their singularities behind the cloak of an event horizon. But even if they do, according to the work for which Hawking is most famous, that cloak may not last forever, and one day the nakedness of the singularity could still be exposed to the Universe at large, with all that that implies.

Black holes are cool

Christodoulou's investigation of the Penrose process and rotating black holes focused not on the energy gained by the particle that escapes from the ergosphere, but on the energy lost by the hole itself. When the hole loses energy to the escaping particle, it also spins more slowly, because it loses angular momentum. You can imagine, and Christodoulou calculated, how another particle might be thrown into the black hole from outside in such a way that it would make the spin of the hole speed up again, giving back the lost angular momentum and adding to the mass energy of the hole. But Christodoulou found that if you do this in the right way to give the hole back exactly the amount of angular momentum it has lost, the extra energy added is always *more* than the energy that was lost when the angular momentum was lost. The change in energy cannot be exactly reversed, provided that we want to reverse, exactly, the change in momentum. This led on to the concept of the irreducible mass of a rotating black hole, which Stephen Hawking related to the area of the Schwarzschild surface around the hole.

Physicists were intrigued by this discovery, because irreversible processes have a special place in nature. They are related to a very important law of physics, known as the second law of thermodynamics; this, put at its simplest, tells us that things wear out. You can see the second law at work if you drop an ice cube into a cup of hot coffee, and watch the ice melt while the coffee cools. Heat flows from the hotter object (coffee) to the colder object (ice cube) until everything is smoothed out into a uniform liquid at the same temperature, in which nothing interesting is going on. The law is related to the perceived flow of time – film the melting ice cube, and if you project the film backwards any audience will know at once that something is wrong. Another way of expressing this law is that the amount of information in the Universe (or in any 'closed system' like an ice cube in a cup of coffee contained in a perfectly insulating box) always decreases. There is more information in the system when it consists of coffee *and* ice, because it is more complex than a system that only consists of lukewarm coffee. Physicists actually measure information backwards – they measure disorder, not order, and call a *loss* of information a *gain* in a property called entropy (equivalent to a gain in disorder). What happens to the average teenager's room if his mother is not allowed in to tidy it up is a good example of increasing entropy. The law that entropy can only increase (or at best stay the same) is a key feature in our understanding of the behaviour of the Universe. So when physicists realized that black holes also possessed a property which could only increase (or at best stay the same), they were intrigued.

In 1971 Hawking showed that a black hole does not have to be rotating for this irreversibility to show up in its behaviour. Even a non-rotating (stationary) black hole has a surface area which can only stay the same (if the black hole does not absorb energy or mass) or increase (if it does swallow up matter or energy). And he showed that if two black holes collide with one another and merge, the area of the event horizon around the new, larger black hole will always be larger than the areas of the two original black holes added together. All this was established by about the time that Uhuru was launched. The analogy between the ever-increasing surface area of a black hole and the ever-increasing entropy of the Universe led Hawking and his colleagues James Bardeen (then at Yale University) and Brandon Carter (who was working alongside Hawking in Cambridge) to develop other analogies between the laws of thermodynamics and the properties of black holes; but at first these were regarded as nothing more than mathematical tricks, with no real, physical significance. There seemed to be an insurmountable problem with saying that the surface area of a black hole *is* a measure of its entropy, because the entropy of a system also provides a measure of its temperature. If black holes had temperatures, then they would have to radiate energy appropriate to their temperatures. And in 1973 'everybody knew' that a black hole couldn't radiate anything at all.

Well, almost everybody. As Hawking has acknowledged (for example, in his book *A Brief History of Time*), the work on 'black hole thermodynamics' which he carried out with Bardeen and Carter was largely stimulated by a desire to prove that one person, who had been claiming that black holes could have a real temperature, was wrong. In their paper,* the three researchers stressed that the analogy between the area of the event horizon and entropy was, in their view, nothing more than an analogy. Although they pointed out that a certain property derived from the area 'is analogous to temperature in the same way that [the area] is analogous to entropy', they emphasized that this property and the area itself 'are distinct from the temperature and entropy of the

---

black hole'. Dogmatically, they went on: 'in fact the effective temperature of a black hole is absolute zero.' But they were wrong.

The person who had challenged, and continued to challenge, what 'everybody knew' was Jacob Bekenstein. At the time he first made the challenge, he was a graduate student at Princeton, working under the supervision of John Wheeler. In his book *A Journey into Gravity and Spacetime*, Wheeler recounts how he inadvertently set Bekenstein going along the path that was to lead to one of the most surprising discoveries about the nature of black holes. Wheeler says that one afternoon in 1970 he and Bekenstein were discussing black hole physics in Wheeler's office at Princeton. Wheeler mentioned, tongue in cheek, the guilt he always felt if he allowed a hot cup of tea to exchange energy with a cold cup of tea, to produce two lukewarm cups of tea. Without changing the total energy of the Universe, such an action increases the amount of disorder, or entropy. Information has been lost forever, a crime which echoes 'down to the end of time', as Wheeler put it. But, he went on, 'if a black hole swims by, and I drop the teacups into it, I conceal from all the world the evidence of my crime.' The point he was making relates to the idea that black holes have no 'hair'. The only properties a black hole has are mass, charge and spin; there is no information about whether it is made of cups of tea or star stuff, or whether the cups of tea dropped into it were hot, cold or lukewarm. The entropy in the tea has gone into the hole, along with the tea itself.

Bekenstein went away, and thought carefully about this half-facetious remark. A couple of days later, he came back to Wheeler with a response. 'You don't destroy entropy when you drop those teacups into the black hole. The black hole already has entropy, and you only increase it!'

With a confidence perhaps related to his inexperience in research, Bekenstein went on to suggest that the area of the event horizon around a black hole really does provide a direct measure of both its entropy and its temperature, and he calculated that a black hole with a mass three times that of our Sun (the smallest black hole that would form from a collapsed star) should have a temperature of rather less than one millionth of a degree above the absolute zero of temperature, $-273°C$, the temperature at which all thermal motion of atoms and molecules stops. Now, this is a very modest temperature, and the calculations require that more massive black holes would have even lower temperatures. But it definitely is not zero, and it implies that somehow energy can leak out of a black hole. All of this appeared in Bekenstein's Ph.D. thesis in 1972, although of course it had been aired earlier in incomplete form.

Hawking, by now something of a black hole expert, was infuriated by Bekenstein's suggestion, which he regarded as complete nonsense. The work with Bardeen and Carter was a direct response to Bekenstein's thesis. Bekenstein was discomfited by the opposition to his arguments (not just from Hawking, but from other researchers, including Werner Israel), and although he persisted in promoting the idea that the area of a black hole was a measure of its entropy, in a paper published in 1973* he echoed the comments of Hawking and his colleagues, saying that the property he had discovered should not be regarded 'as *the* temperature of the black hole: such an identification can easily lead to all sorts of paradoxes, and is thus not useful.' Yet, even as Bekenstein's faith in his own judgement seemed to be wavering, his theory was soon to gain support from an unexpected source.

Also in 1973 Hawking learned, on a visit to Moscow, that two Soviet researchers, Yakov Zel'dovich and Alex Starobinsky, had discovered that rotating black holes could create particles out of energy and eject them into space. This was an interesting and acceptable idea, since the energy to make the particles could come from the ergosphere in a kind of Penrose process. But when Hawking tried to develop a proper mathematical treatment of the Zel'dovich-Starobinsky effect, using quantum mechanics, he found to his surprise and initial annoyance that the equations said that even non-rotating black holes would emit particles. He had arrived, by a different route and against his own inclinations, at the same conclusion as Bekenstein. Black holes *do* have a temperature, he acknowledged in 1974, and they *do* emit particles – a phenomenon now known as the Hawking process (which seems a little unfair on Bekenstein,† Zel'dovich and Starobinsky). But the temperature of the hole is not an independent property to add to mass, spin and charge – the temperature depends on the area of the event horizon, which is itself already determined from the three basic properties. Even a hot black hole still has no hair.

The simplest way to understand the Hawking process is in terms

---

* *Physical Review*, Volume D7, pp. 2333–46.

† Adding insult to injury, in both the 1973 paper arguing that Bekenstein was wrong and the 1974 paper acknowledging that he was right (*Nature*, Volume 248, pp. 30–31), the name of Hawking's protagonist was mis-spelled 'Beckenstein'.

of a combination of quantum physics and relativity theory. Relativity tells us that energy can be converted into matter. Quantum physics tells us that there is always an intrinsic uncertainty in the amount of energy in a system. Among other things, this means that no system can ever have precisely zero energy – if it did, there would be no uncertainty. Even 'empty space' contains energy, which cannot be measured directly, but which can create short-lived pairs of particles, which flicker in and out of existence in an incredibly short timescale, less than 10^{-44} of a second. The particles have to come in pairs, in order to ensure that there is always a balance of quantum properties such as electric charge. So, for example, every temporary electron that is formed this way (carrying negative charge) is paired with a temporary positron (carrying positive charge).* Such 'particle-antiparticle pairs' almost immediately annihilate one another, giving back to the vacuum the energy they have temporarily borrowed. They are known as 'virtual' pairs (Figure 5.6). Although the concept sounds bizarre, the presence of this sea of virtual particles has a measurable effect on the behaviour of real, charged particles, and no physicists doubt their existence.

But what happens to virtual pairs produced right on the edge of

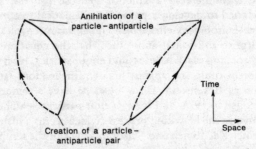

Figure 5.6 *What we think of as 'empty' spacetime is actually full of a seething ferment of 'virtual' particles, which are created in pairs out of nothing at all, by quantum uncertainty, but promptly annihilate one another, forming closed world-line loops.*

* For an explanation of all the quantum rules, see my book *In Search of Schrödinger's Cat*.

the horizon of a black hole? In a process reminiscent of the Penrose process, one member of the pair can cross the horizon and be swallowed up by the hole, leaving the other particle with no partner to annihilate with. The leftover particle gains energy from the gravitational field of the hole, becoming a real particle that scoots away out into the Universe (Figure 5.7). Just as in the case of the

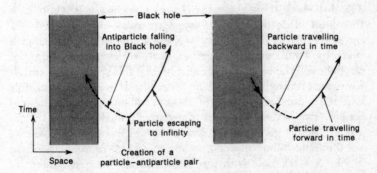

Figure 5.7 *If a virtual pair of particles forms next to a black hole, one member of the pair may fall into the hole, leaving the other one with nothing to annihilate with. It is 'promoted' into reality, gaining energy from the mass of the black hole itself. The process is exactly the same, as far as the equations are concerned, as if a particle tunnelled out of the hole by travelling backwards in time, and then scooted off into the future once it was at a safe distance from the event horizon.*

Penrose process, in the Hawking process the black hole itself loses mass energy and its surface area shrinks. It is the energy carried by particles evaporating in this way from all over the event horizon (now known as Hawking radiation) that provides the temperature that Bekenstein had suggested a black hole must have. The temperature is related to the area of the event horizon in such a way that the bigger a black hole is, the lower its temperature is.

For stellar mass, and larger, black holes, it is a slight exaggeration to say that 'black holes are hot' (indeed, it is a slight exaggeration to say that black holes are cool!), and if this was the end of the story of Hawking radiation, the process would actually be a welcome addition to our understanding of black holes. The process does, after all, unite the three great theories of physics – thermodynamics, relativity and quantum mechanics – within one framework. But there is another, more disturbing, aspect to Hawking's discovery.

Exploding horizons

The disturbing thing about the Hawking process is that it tells us that *small* black holes might evaporate away entirely, leaving a naked singularity behind. Any large black hole will, of course, gain mass by swallowing particles of dust, or even the energy of starlight and the background radiation that fills space, faster than it loses mass through the Hawking process. But imagine a black hole that starts out with a mass of about a billion (thousand million) tonnes. This is about the mass of a relatively small asteroid, such as Apollo, or a large mountain, such as Mount Everest; it would take six thousand billion objects this size to add up to the mass of the Earth. A black hole with this astronomically modest amount of mass would have a Schwarzschild radius of only about 10^{-13} cm, roughly the size of the nucleus of an atom. It would be very hard for such a black hole to swallow anything at all – even 'eating' a proton or a neutron would be quite a mouthful for it. And yet, according to Hawking's calculations, it would have a temperature of about 120 *billion* degrees, and be radiating away energy furiously at a rate of 6,000 megawatts, equivalent to the output of six large power stations. The flow of positive energy outwards into space would be balanced by a flow of negative energy into the hole, which would shrink as a result. The more it shrank, the hotter it would get, and the faster it would shrink. In the end, such an object might disappear completely, leaving nothing behind but the radiation it has emitted. Alternatively, it may be that quantum effects will stop the radiation when the hole gets down to a certain size. But there is a third possibility – that the emission of radiation might shrink away the event horizon around the black hole, until the horizon vanishes, leaving a naked singularity behind. What's worse, the calculations tell us that this singularity would have negative mass, balancing all the positive energy that had poured out of the hole and into the Universe at large.

This wouldn't be so troubling if we only had to worry about black holes that form in the Universe today. As things are now, you could only make a black hole by starting out with several times more mass than there is in our Sun, and letting it collapse gravitationally. And any black hole formed in this way will only have a tiny temperature, producing a feeble amount of Hawking radiation. But three years before he discovered that mini black holes should explode, Hawking had already suggested that mini black holes might exist. In 1971, he had pointed out that under the extreme conditions of very high pressure that existed in the big bang itself, even black holes with masses as small as one hundred-thousandth of a gram might have been produced. After all, you can make a black hole out of *anything* if you squeeze it hard enough – remember that the Earth itself would become a black hole if we had any means of squeezing it down into a sphere with a radius of just under a centimetre. Unless the big bang itself was perfectly smooth, there must have been irregularities, with some regions slightly more dense than average, and some slightly less dense than average, and it would have been inevitable that some of the 'overdense' regions would come out of the big bang as black holes.

So, in two quite separate pieces of work, Hawking had established that tiny 'primordial' black holes might very well be at large in the Universe, and he had shown that such objects must evaporate away, very probably leaving negative-mass naked singularities behind. I chose the particular example of a mini black hole that starts out with a mass of a billion tonnes because the calculations show that the time that has elapsed since the big bang is just long enough for holes with this much mass (and, of course, any with less mass than this) to have evaporated away by now. All of this was not welcome news to most physicists. Remember how Eddington had ridiculed Chandrasekhar's discovery of ultimate gravitational collapse?

> The star has to go on radiating and radiating, and contracting and contracting, until, I suppose, it gets down to a few km radius, when gravity becomes strong enough to hold in the radiation, and the star can at last find peace.

Hawking had turned that on its head. Paraphrasing Eddington, you might say that what Hawking had shown was that:

> The black hole has to go on radiating and radiating, and contracting and contracting, until, I suppose, the event horizon disappears, and the singularity inside is exposed to the Universe.

And not just any old singularity, but a negative-mass naked singularity, at that! The edge of time itself, exposed for all to see. Even in the 1990s, many physicists find the notion as ridiculous as Eddington found the notion of black holes in the 1930s. But some are made of sterner stuff. In the BBC radio broadcast that I have already mentioned, after pointing out that 'there is no very convincing theoretical argument in favour of Cosmic Censorship', Roger Penrose said

> It has often been argued that if naked singularities arise then this situation would be disastrous for physics. I do not share such feelings. True we have, as yet, no theory which can cope with space-time singularities. But I am an optimist. I believe that eventually such a theory will be found.

In the spirit of Penrose, it is almost time to find out how the existence of such edges to the Universe allow for the possibility of travel through both space and time. But first, I want to step back a little from the singularity, and take a brief look at some of the strange things that go on in the region of spacetime just *outside* the event horizon of a black hole. For, although the strangest and most extreme distortions of spacetime are associated with the singularity itself, even going close by a black hole, and not venturing across the horizon at all, can leave you in something of a whirl – both mentally and physically.

Centrifugal confusion

The first strange thing that happens to you if you travel near a black hole involves the sensation we call centrifugal force. Centrifugal force is familiar to anyone who has ridden in a car speeding round a curve. It is the force that pushes you outwards as you go round the bend. Of course, what really happens is that your body tries to carry on in a straight line, and is pushed sideways by the car seat, the side of the vehicle, or your safety belt. You probably recall your school physics teacher explaining that centrifugal force is a 'fictitious' force, which is simply a result of rotation. But that doesn't mean that the force is not real enough for anybody who is in what is known as a rotating frame of reference. If you place a tennis ball on the dash of the car, then when the car turns to the right the ball rolls to the left, outward from the curve. In the frame of reference tied to the car, there *is* a force pushing the ball outward. Apart from the little quibble about whether this should be called a fictitious force or not, we all know what is going on, and which way the ball will roll. You would be amazed if the car turned sharply to the right, and the ball, in response, rolled smartly to the right across the dash. But according to Marek Abramowicz, working in 1990 at NORDITA (the Scandinavian theoretical physics institute in Copenhagen), that is exactly what would happen if your car was a spaceship, and the sharp right turn you were making was taking you skimming above the event horizon of a black hole.

Abramowicz has been puzzling since the early 1970s over some curious predictions concerning centrifugal force that emerge from the equations of general relativity. In 1990 he and his colleagues established that centrifugal force actually works the other way, squeezing you against the *inner* side of a vehicle following a circular path, if that path is an orbit skimming the surface of a black hole. This happens only for orbits that pass within a certain distance from the event horizon, and that distance is related to another feature of black holes that often causes confusion.

Recall that the surface of the event horizon lies at the distance from the central singularity where the escape velocity is equal to the speed of light. If you had a rocket with an infinitely powerful motor and an infinite supply of fuel, then by pointing the rocket exhaust towards the singularity and blasting the motor at full power you could just hover, motionless, on the event horizon. But the event horizon is *not* at the distance from the singularity where light rays are bent into a circle around the singularity. That actually happens a little further out, at a distance from the singularity one and a half times the Schwarzschild radius. Between the Schwarzschild radius and this distance, which defines the 'speed-of-light circle', a light ray cannot stay in orbit around the black hole. Any light ray that comes within the speed-of-light circle must either plunge into the black hole itself or be bent around the hole and emerge from its vicinity, following an open curve out into space again. Between the event horizon and the speed-of-light circle, your infinitely powerful rocket ship could balance the force of gravity at any distance from the hole by judicious use of its rocket motors. Then, with the aid of a sideways-pointing rocket, it could travel in a circular orbit around the hole. This is where the fun begins.

In fact, the fun begins at the speed-of-light circle. For the circular photon orbits themselves, centrifugal force is zero – and it reverses direction as you cross those orbits. Abramowicz explains this in physical terms (the mathematical treatment takes rather longer) by pointing out that the path of a light ray defines a geodesic, the relativistic equivalent of a straight line. Since centrifugal force acts only when we move along a curved path, anything that moves along a light path cannot feel centrifugal acceleration. And this applies for *any* spaceship, orbiting the black hole at *any* speed, provided that it follows the circular path of a trapped photon. Provided that the rocket motor balances the pull of gravity to keep the spaceship at exactly the distance of the speed-of-light circle from the black hole, the sideways rocket can push the spaceship around that circle at any speed at all, and the occupants of the spaceship will be weightless, in free fall, and will feel no centrifugal force at all.

This is quite different from the weightlessness experienced by astronauts falling freely in orbit around the Earth. There, the astronauts and their spacecraft are simply falling in a natural orbit under the influence of gravity. But our hypothetical black hole explorers are forcing their spacecraft into an unnatural orbit by firing its motors continuously. And yet, they are still weightless.

For other circular orbits, there will be just one speed where centrifugal force balances gravity, and where the spaceship will stay in orbit without firing its motors, just like a spaceship orbiting the Earth. In such an orbit, the occupants of the spacecraft will be weightless. For all other orbital speeds, in order to maintain the same distance from the central mass the rocket motors must be used continuously to push the spaceship in or out with the appropriate force. In those circumstances, the occupants will feel a centrifugal force pushing them against one of the walls of the spacecraft. But in these special photon orbits, the motors only have to be used to balance the gravitational pull of the central mass. Once that is done, the spacecraft can glide around the circular orbit at any speed at all, in free fall.

Within these circular photon orbits, though, centrifugal force *adds* to the inward pull of gravity. So the outward force needed to keep the spaceship in a circular orbit increases as the speed of the spaceship round that orbit increases. Instead of being flung outward by centrifugal force, the occupants of the fast-moving spacecraft are sucked inward. In other words, centrifugal force always acts in such a way as to repel orbiting particles from the circular photon orbits.

All this is of more than just esoteric interest to relativists. The only black holes that have yet been identified in the Universe are those like Cygnus X-1 where matter is being swallowed up by the hole as it is torn from a companion star by tidal forces. This infalling matter forms a swirling accretion disk, in which very high temperatures are reached and X-radiation is produced. It is the X-rays that reveal the presence of the black hole to observers on Earth.

But how does the accretion disk feed matter into the hole? According to the new insight provided by Abramowicz and his colleagues, once the material crosses the region of circular photon orbits, rotation will force it right into the hole, no matter how fast it orbits. It is as if you swirled the tea in your cup, and instead of piling up at the outside to form a concave surface, the swirling tea piled up in the middle, to make a hump. Processes like this at work in the accretion disk around a black hole will affect the production of X-rays by the source, and so future observations could reveal the effects of the reversal of centrifugal force, even without intrepid astronauts flinging themselves into close orbit around a black hole to carry out the appropriate measurements.

But centrifugal reversal isn't the only strange thing that such spacefarers would be able to measure. Pick a big enough black hole, where tidal forces near the event horizon are not too extreme, and, still without ever crossing the event horizon, they would be able to use the region of distorted spacetime around the hole for repeated journeys through time – but only in one direction, into the future.

A one-way time machine

The role of gravity in slowing the flow of time near a black hole is not in doubt. This is simply a more extreme version of distortions in spacetime that have already been measured, particularly in terms of the gravitational red shift of light from white dwarf stars.

Previously, I described the gravitational red shift in terms of the energy lost by light struggling out of the gravitational well of a very dense object. But the gravitational time dilation effect gives us another perspective on what is going on. From this point of view, light itself can be used as a clock. Because light travels at a steady speed, 300,000 km a second, light with a particular wavelength can be used to measure the passage of time. The electromagnetic waves that make up light are, as Maxwell pointed out, oscillating electric and magnetic fields. If we choose one of the two components, for simplicity, such a wave can be represented, moving through space, as a wiggly line like the one shown in Figure 5.8. The amplitude of the wave measures the size of the oscillation, and the wavelength is the distance from one peak of the oscillation to the next. If you now imagine watching the wave go by, and simply count the successive peaks as they pass, you will see that the time between each peak is simply the wavelength divided by the speed of light. Each wave peak can be thought of as a little flicker of energy, and for light with a particular wavelength (a particular colour) the flickers follow each other at regularly spaced time intervals, like the ticking of a perfect clock. In fact, this is how the length of time we call the second is defined.

The second was originally defined in terms of the rotation of the Earth, which is the basic astronomical 'clock'. There are sixty seconds in a minute, sixty minutes in an hour, and twenty-four hours in a day, so the length of the second was defined as 1/86,400 times the length of the day. But the length of the day changes slightly during the course of the year, as the Earth moves around the Sun. On longer timescales, the spin of the Earth is gradually slowing down; and there are also more erratic changes, for example, when our planet is shaken by large earthquakes. All in all, the spinning Earth is a far from perfect timekeeper. So the second is now defined in terms of the frequency of a particular pure wavelength of radiation emitted by atoms of caesium – one second is the time it takes for this specific electromagnetic wave to flicker 9,192,631,770 times. This definition of time is what gives us the expression 'atomic clocks'; in fact, though, atomic clocks are really light clocks. All our time signals come ultimately from such clocks

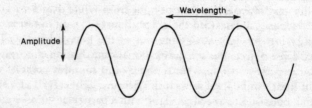

Figure 5.8 *A wave.*

today, and when you set your watch by the radio time signal you are matching it up to the 'ticking' of light from caesium atoms. But the Earth itself continues to be an erratic timekeeper, and we want our everyday clocks and watches to continue to show noon when the Sun is at its highest in the sky. So as the Earth slips slightly out of step from atomic time, occasionally the official radio time signals include allowance for an extra 'leap second' to bring things back in line, ensuring that noon on your watch is never more than one second away from the time when the Sun is highest in the sky (assuming you have set your watch correctly, and that it keeps perfect time). What matters for the purpose of discussing red shift, however, is that each second is precisely the same length as any other second, and is defined in terms of the frequency of vibration of electromagnetic waves – light.

Now apply this to measurements made near a black hole. Astronauts travelling close to a black hole could carry a caesium clock with them. They would measure the wavelength of the all-important electromagnetic radiation from the atoms, and find that it was exactly the same as back home on Earth. They would happily set their clocks by the vibrations of this wavelength of light, and proceed about their business as usual. But if the radiation from the caesium atoms was beamed out from the vicinity of the black hole and picked up by observers out in the region of flat spacetime, they would find that it had a longer wavelength, because of the gravitational red shift, than equivalent radiation from identical atoms on Earth, or from caesium atomic clocks in the observers' spacecraft. In other words, the time between successive flickers of energy would have increased in the red-shifted light. In the time it would take for 9,192,631,770 flickers from the atomic clock down near the black hole to pass by the observers out in flat spacetime, much more time than one second would have passed according to the clocks out in the region of flat spacetime. Compared with events in flat spacetime, the astronauts in the region of strong gravitational field near the black hole would be living more slowly.

But to them, of course, everything on board their spacecraft would seem to be normal. Indeed, *they* would argue that the observers out in flat spacetime were living speeded-up lives! After all, if light from the caesium atoms of those observers' clocks were beamed down into the region just above the black hole, it would gain energy from the gravitational field and be blue-shifted to higher energy and therefore shorter wavelengths, corresponding to higher frequency. Comparing this incoming light with the radiation from their own clocks, the astronauts would conclude that time was running fast in the outside Universe.

Either perspective would be correct. If the astronauts now fired their rockets to come out of the region of highly distorted spacetime to compare clocks with the observers, the clocks that had travelled down near the black hole with the astronauts would show that less time had elapsed, while the clocks that had stayed out in flat spacetime with the observers would show that more time had elapsed. What's more, the astronauts would literally have aged less than the observers. This whole time dilation business is *not* some illusion caused by the way we choose to measure time. Choosing to measure time from the spin of the Earth has no significance for the Universe at large; but choosing to measure time in terms of the oscillations of electromagnetic waves is indeed truly fundamental. It is the way the Universe itself measures time. Light, as Einstein realized, provides the only, and essential, infallible, fundamental measure of both length and time in the Universe. If you find it hard to accept that the astronauts who visit the region just above the event horizon will have aged less than the observers who stayed in flat spacetime, remember that those astronauts and observers are themselves made of atoms. If gravity affects the way caesium atoms produce light, it should be no surprise to find that gravity affects the way atoms in living bodies behave. Time really does run slow in the region of spacetime near a black hole.

And this is what makes it possible to use a comfortably large black hole as a one-way time machine. The longer the astronauts spend near the event horizon, and the closer they get to it, the stronger the effect will be. You don't even need enormously powerful rockets to take advantage of the effect, because the astronauts could use a judicious, short-lived blast on their rockets to set their spacecraft falling on an open orbit down into the region of highly distorted spacetime, leaving the observers behind in a space station orbiting in a circular orbit far from the black hole. The falling spacecraft would coast in, in free fall all the way, being accelerated by the gravity of the black hole up to the point of closest approach. Then, it would whip round the hole very sharply (still in free fall, but perhaps producing some vertiginous tidal effects) and climb out again, now being slowed all the time by gravity. At the furthest distance from the hole, the astronauts could fire their rockets briefly again, to put the spacecraft back alongside the space station of the observers who had stayed out in a high orbit, ready to compare clocks. By choosing the right path around the black hole, such a journey, which might take a few hours according to the clocks on the falling spacecraft, could be made to take as long as you like according to the outside Universe. A hundred years, a thousand years, or longer. What's more, the astronauts could repeat the process as often as they wanted to, hopping down the centuries and millennia into the future. On such a scenario, the observers they met on each visit into the region of flat spacetime would not be the original observers who watched them set off. Those observers would long since have died of old age, and been replaced by successive generations of new observers.

It sounds like science fiction, and the basic idea has indeed been used in more than one science fiction story. But it is all sober scientific fact. The one snag with the scenario (which is what makes those stories science *fiction*) is that in order to take advantage of the one-way time machine process you would first have to find, and then travel to, a very massive black hole, in order to avoid tidal problems. The nearest black hole likely to be large enough to fit the bill is the one at the centre of the Milky Way, more than 30,000 light years away. Even light takes more than thirty thousand years to reach us from the vicinity of the nearest usable one-way time machine. Since nothing can get there faster than light, in order to take advantage of the possibilities it offers, we would have to find a shortcut through space so that we could get to it in reasonable time. That, of course, is another familiar science fiction device, as well – the concept of tunnels through 'hyperspace'. Only, would you believe, that idea too turns out to be based on respectable scientific fact.

CHAPTER SIX

Hyperspace Connections

When *science fiction becomes fact. White holes, wormholes and spacetime tunnels. Journeys into other universes, and our own past. The blue sheet block – and a way around it. Opening the throat of hyperspace, with the aid of antigravitational string*

When astronomer Carl Sagan decided to write a science fiction novel, he needed a fictional device that would allow his characters to travel great distances across the Universe. He knew, of course, that it is impossible to travel faster than light; and he also knew that there was a common convention in science fiction that allowed writers to use the gimmick of a shortcut through 'hyperspace' as a means around this problem. But, being a scientist, Sagan wanted something that would seem to be more substantial than a conventional gimmick for his story. Was there any way to dress up the mumbo-jumbo of Sf hyperspace in a cloak of respectable-sounding science? Sagan didn't know. He isn't an expert on black holes and general relativity – his background specialty is planetary studies. But he knew just the person to turn to for some advice on how to make the obviously impossible idea of hyperspace connections through spacetime sound a bit more scientifically plausible in his book *Contact*.

The man Sagan turned to for advice, in the summer of 1985, was Kip Thorne, at CalTech. Thorne was sufficiently intrigued to set two of his Ph.D. students, Michael Morris and Ulvi Yurtsever, the task of working out some details of the physical behaviour of what the relativists know as 'wormholes'. At that time, in the mid-1980s, relativists had long been aware that the equations of the general theory provided for the possibility of such hyperspace connections. Indeed, they are an integral part of the Schwarzschild solution to Einstein's equations, and Einstein himself, working at Princeton with Nathan Rosen in the 1930s, had discovered that Schwarzschild's solution actually represents a black hole as a bridge between *two* regions of flat spacetime – an 'Einstein-Rosen bridge'. This is related to the fact that there are two sets of solutions to the equations, mentioned in Chapter Five. But before Sagan set the ball rolling again, it had seemed that such hyperspace connections had no physical significance and could never, even in principle, be used as shortcuts to travel from one part of the Universe to another.

Morris and Yurtsever found that this widely held belief was wrong. By starting out from the mathematical end of the problem, they constructed a spacetime geometry that matched Sagan's requirement of a wormhole that could be physically traversed by human beings. *Then* they investigated the physics, to see if there was any way in which the known laws of physics could conspire to produce the required geometry. To their own surprise, and the delight of Sagan, they found that there is. (Almost certainly, Thorne was not surprised by the discovery; Morris recalls that when Thorne set his students the problem, he had a distinct impression that Thorne himself had already figured out the answer, although it took Morris a few weeks to follow through the hints he gave them.) To be sure, the physical requirements seem rather contrived and implausible. But that isn't the point. What matters is that it seems that there is nothing in the laws of physics that forbids travel through wormholes. The science fiction writers were right – hyperspace connections do, at least in theory, provide a means to travel to far distant regions of the Universe without spending thousands of years pottering along through ordinary flat space at less than the speed of light.

The conclusions reached by the CalTech team duly appeared as the scientifically accurate window-dressing in Sagan's novel when it was published in 1986, although few readers can have appreciated that most of the 'mumbo-jumbo' was soundly based on the latest discoveries made by mathematical relativists. Since then, the discovery of equations that describe physically permissible, traversable wormholes has led to a booming cottage industry of mathematicians investigating these strange phenomena. It all starts with the Einstein-Rosen bridge; and before we can see just how startling the discoveries made by Thorne and his students were, we need to take stock of the conventional wisdom which said, when Sagan posed his question in 1985, that wormholes were figments of the mathematics that had no physical significance at all.

The Einstein connection

It's one of the intriguing curiosities of the history of science – and a good example of how it is virtually impossible to tell the story of black holes without jumping about in the historical narrative – that spacetime wormholes were actually investigated by mathematical relativists in great detail long before anybody took the notion of black holes seriously. As early as 1916, still less than a year after Einstein had formulated his equations of the general theory, the Austrian Ludwig Flamm had realized that Schwarzschild's solution to Einstein's equations actually describes a wormhole connecting *two* regions of flat spacetime – two universes. Speculation about the nature of wormholes continued intermittently for decades; the most notable contributors to the discussion were Hermann Weyl in the 1920s (Weyl was a German mathematician who studied at Göttingen, Riemann's old home, and specialized in investigating Riemannian geometry), Einstein and Rosen in the mid-1930s, and John Wheeler in the 1950s. But their interest was not in the kind of large, traversable wormholes that became the stuff of science fiction (known as 'macroscopic' wormholes).

People first became interested in wormholes through thinking about the nature of fundamental particles, like electrons. If an electron existed literally as a point of matter, then the correct way to describe spacetime around that point would be using the Schwarzschild metric, complete with a tiny wormhole (known, for obvious reasons, as a 'microscopic' wormhole) providing a connection to another universe. The theorists I have mentioned, and others, wondered whether all fundamental particles might actually be microscopic wormholes, and whether properties such as electric charge might arise because fields of force (in this case, the electric field) were threaded through the wormholes from the other universe. Such ideas had obvious appeal for Einstein and other relativists, since they raised the possibility of explaining the structure of matter in terms of particles which are ultimately the products of warped spacetime – in other words, explaining *everything* in terms of general relativity. Their hopes were not fulfilled, although (as we shall see in Chapter Eight) an intriguing variation on this theme is once again causing a flurry of interest among the relativists in the 1990s. What the pioneering relativists did establish, very early on, was that Schwarzschild wormholes provide no means of communicating from one universe to the other.

The problem is easily understood, in its modern form, in terms of a Penrose diagram, like Figure 6.1. A Schwarzschild wormhole, or Einstein–Rosen bridge, linking the two universes can be represented by a line drawn across the page, linking the two sides of the diagram. But remember that the diagonal lines on such a diagram correspond to motion at the speed of light, and that lines making shallower angles correspond to motion faster than the speed of light. In order to traverse an Einstein–Rosen bridge from one universe to the other, a traveller would have to move faster than light at some stage of the journey, whatever kind of wiggly line you might draw linking the two universes. And there is another problem with this kind of wormhole – it is unstable. If you imagine the 'dent' in spacetime made by a large mass such as the Sun, squeezed into a volume only slightly bigger than its corresponding

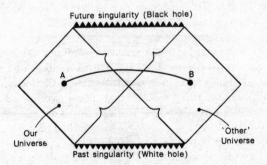

Figure 6.1 In order to travel from one universe to the other, you would have to travel 'across the page'. This involves motion at an angle shallower than 45°, which means faster than light. At first sight, it seems impossible.

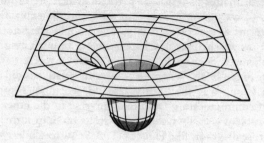

Figure 6.2 A reminder of the 'embedding diagram' that represents the way an object like the Sun distorts spacetime.

Schwarzschild sphere, you would get an 'embedding diagram', like Figure 6.2. The surprise about the Schwarzschild geometry is that when you shrink the mass down to within its Schwarzschild radius, you don't just get a bottomless pit, as in Figure 6.3; instead, the bottom of the embedding diagram opens out to make the connection with another region of flat spacetime (Figure 6.4). But this beautiful, open throat, offering the tantalizing prospect of travel between universes, exists for only a tiny fraction of a second. If we look again at a Penrose-type diagram, we can take slices through the diagram corresponding to different times ('past' is at the bottom

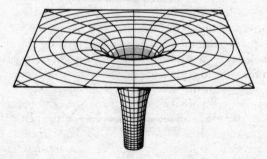

Figure 6.3 We usually think of a black hole as an extreme version of an embedding diagram, with, literally, a hole in the structure of spacetime.

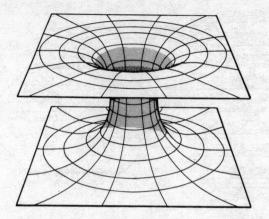

Figure 6.4 *But remember that there is an extra version of spacetime built into the equations. A black hole is really a throat – or 'wormhole' – connecting two universes.*

of the diagram, while 'future' lies at the top), and draw an embedding diagram corresponding to each slice of space (Figure 6.5). This shows that the Schwarzschild throat forms from two distortions in the opposite regions of flat space that grow towards each other, connect, and open out to full size, then shrink away again, disconnect, and separate. For a black hole with the same mass as our Sun, the entire evolution of the wormhole, from the disconnected state associated with the singularity in the past, through the Schwarzschild throat state and on to the disconnected state corresponding to the singularity in the future, takes less than one ten-thousandth of a second, as measured by clocks inside the black hole. The wormhole itself does not even exist for long enough for light to cross from one universe to the other. In effect, gravity slams shut the door between universes.

This is especially disappointing, because if you ignore the rapid evolution of the wormhole and look only at the geometry corresponding to the instant when the throat is wide open, it seems as if such wormholes might even connect, not separate *universes* but separate regions of our *own* Universe. Space may be flat near each mouth of the wormhole, but bent around in a gentle curve, far away from the wormhole, so that the connection really is a shortcut from one part of the Universe to another (Figure 6.6). If you

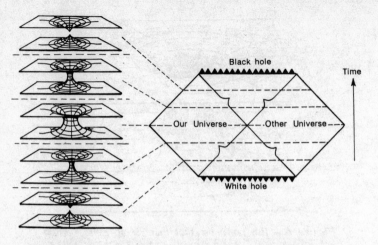

Figure 6.5 *The snag is that the situation shown in Figure 6.4 is really only a momentary snapshot of the wormhole. In fact, as time passes the wormhole opens up and then snaps shut again. It does so too quickly for anything – even light – to pass through.*

imagine unfolding this geometry to make the entire Universe flat except in the vicinity of the wormhole mouths, you get something like Figure 6.7, in which a curved wormhole connects two separate regions of a completely flat Universe – and don't be fooled by the

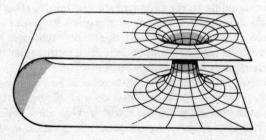

Figure 6.6 *If a way could be found to hold a wormhole open, it might even provide a shortcut from one region of our Universe to another.*

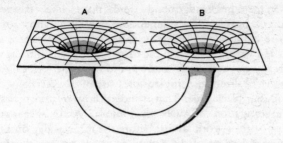

Figure 6.7 *The wormhole is still a shortcut in four dimensions even when the way we draw it on paper makes it look as if it is the long way round.*

fact that in this drawing the distance from one mouth to the other through the wormhole itself seems to be longer than the distance from one mouth to the other through ordinary space; in the proper four-dimensional treatment, even such a curved wormhole can still provide a shortcut from A to B.

Or at least, it could if the wormhole stayed open for long enough, and if passage through the wormhole didn't involve travelling at speeds faster than that of light. The second problem is a direct result of the fact that the future singularity in the Penrose diagram of a Schwarzschild black hole lies horizontally across the page, so that anything (or anybody) that crosses the event horizon has no option but to crash into the singularity. But this is not the end of the story of hyperspace connections. A simple Schwarzschild black hole has no overall electric charge, and it does not rotate. Intriguingly, adding *either* electric charge *or* rotation to a black hole transforms the nature of the singularity, thereby opening the gateway to other universes, *and* makes the journey possible while travelling at speeds less than that of light.

Charging through hyperspace

Nobody thinks it very likely that electrically charged black holes exist. If a black hole somehow built up a large positive charge, for example, it would soon neutralize itself by swallowing negatively charged particles (such as electrons) from its surroundings, while repelling any additional positively charged particles that came its way. On the other hand, nobody thinks it very likely that real black holes do *not* have angular momentum; they surely must rotate, and the more compact they are, the faster we would expect them to rotate. However, it is simplest to get an insight into how black holes really might provide gateways to other universes by looking first at the idealized, and implausible, case of an electrically charged, non-rotating black hole. That is indeed how relativists first began their investigation of such phenomena. And, once again, the pioneers were at work almost before the ink was dry on Einstein's definitive statement of the equations of the general theory, while the First World War was still raging across Europe.

The description of the spacetime structure near a charged (but non-rotating) black hole is known as the Reissner-Nordstrøm geometry – but, unlike Einstein and Rosen, Reissner and Nordstrøm did not work together. Heinrich Reissner, in Germany, was first off the mark, publishing a paper on the self-gravitation of electric fields in the context of Einstein's theory in 1916; the Finn Gunnar Nordstrøm added his contribution in 1918. But although they did not work together, their names are now permanently linked in the pages of text-books on relativity. Once again, it is easiest to understand the importance of their discoveries by thinking in terms of the standard relativists' visual aid, the Penrose diagram.

Adding electric charge to a black hole provides it with a second field of force, in addition to gravity. But because charges with the same sign (all positive, or all negative) repel one another, this electric field acts in the opposite sense to gravity, trying to blow the black hole apart, not pulling it more tightly together. Of course, nothing can make the black hole explode outwards (unless, and until, Hawking radiation has shrivelled the event horizon itself away to nothing at all). But this does still mean that there is a force inside a charged black hole which, in a sense, opposes the inward tug of gravity. The most important consequence of this is that there is a second event horizon associated with a charged black hole, inside the event horizon associated with the hole's gravitational field.

What this means in physical terms is that there are two spherical surfaces surrounding the central singularity, one inside the other, both marking locations where time, as measured by a distant observer, comes to a standstill. The outer event horizon is a little closer to the singularity than the event horizon for a black hole with the same mass but no electric charge; the inner event horizon is close to the singularity if the hole has only a small amount of electric charge, and further away if the charge is greater. In principle, if the black hole had enough electric charge the inner horizon would move out *past* the outer event horizon; then, both horizons would vanish and we would be left with a naked singularity. But this would require adding a vast electric charge to a black hole, and there is no conceivable practicable way in which this could be done. Nevertheless, the Reissner-Nordstrøm equations seem to take no notice of the principle of cosmic censorship. Even stranger, an astronaut who boldly took a flight close up to such a singularity would not be attracted towards it and crushed by gravity – the Reissner-Nordstrøm singularity actually *repels* objects that approach it too closely, acting as a region of antigravity.

But even this is only the beginning of the weirdness of charged black holes. Remember that as you cross the event horizon of a Schwarzschild black hole the roles of space and time are reversed. As a result, the world line of the singularity in a Penrose diagram is not that of a point in space, travelling vertically 'up the page' as time passes. Instead, once you cross the horizon on the way into a black hole the singularity stretches in front of you across all of space, and you must inevitably fall into it. If you were to fall into a Reissner-Nordstrøm black hole, however, there would be a *second* reversal of the roles of space and time, as you crossed the *second* event horizon. As a result, the representation of the singularity in the Penrose diagram is not as a horizontal bar across the page, but as a vertical line, up the page (Figure 6.8). By careful navigation of your spaceship, you could avoid the singularity, even though always travelling at speeds less than that of light, and cross the event horizons once again on your way out of the black hole! Although gravity still tries to slam shut the door opening to other universes, the electric field holds the door open for travellers to get through. But there is still a sense in which this is a one-way door; you could not get back to the universe you started from – the one-way nature of the event horizons depicted in Figure 6.8 means that you would inevitably emerge into another region of space-time, usually interpreted as another universe. Turning around to go back the way you came *would*, as these new maps of spacetime make clear, require travelling faster than light. Even this is not the end of the story. Look again at Figure 6.8. The spacetime map is open-ended. Instead of just two universes connected by the

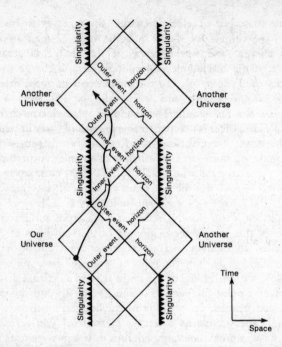

Figure 6.8 *The spacetime map of an electrically charged black hole shows how the hole connects many universes (or many regions of one Universe) together. Because the singularity is now vertical, an expert navigator in a suitable spaceship could steer a course through the black hole and into a different region of flat spacetime, without ever having to exceed the speed of light.*

Schwarzschild geometry, the Reissner-Nordstrøm geometry connects an infinite number of such pairs of universes in a chain. This kind of spacetime map is sometimes referred to as the 'paper doll' topology, because the repeating pattern resembles the chain of linked paper dolls you can make by cutting a folded sheet of paper – but each paper doll is an entire *pair* of universes (Figure 6.9).

This is all very well in principle, but since charged black holes almost certainly do not exist in our Universe, it is of no more than esoteric interest. Except for one thing. The effect of rotation on the spacetime geometry of a black hole is similar, in some ways, to the effect of electric charge. In particular, the angular momentum of a

Figure 6.9 *The spacetime map for a set of universes connected by a charged black hole is like a never-ending chain of paper dolls.*

rotating black hole also opposes the inward tug of gravity, and pushes an inner event horizon out from the singularity, opening the door to other universes. Unlike charged black holes, rotating black holes are certain to exist. And a rotating black hole has another peculiarity all its own – the singularity in the centre is not a mathematical point, but a ring, through which (if the hole is massive enough and is rotating fast enough) a daring space traveller might even dive, and live to tell the tale. Until Sagan made his innocent enquiry about wormholes to Thorne, this was the nearest the mathematicians had come to describing a plausible traversable, macroscopic wormhole.

Bridging the universes

Like the boundary of the ergosphere of a Kerr black hole, the inner event horizon of a rotating black hole bulges out furthest from the centre of the hole around the equator, and doesn't bulge out at all at the poles. This complicates the geometry of spacetime around a Kerr black hole, and helps to explain why it took mathematicians so long to solve the relevant equations – the Reissner-Nordstrøm variation on the black hole theme is spherically symmetric (the same in all directions), and that almost always makes the equations easier to solve. Once Kerr had found out (in 1963) how to allow for the effects of rotation, it was relatively straightforward to add the effects of electric charge in as well. This was done by Ezra Newman and colleagues at the University of Pittsburgh in 1965; their solution to Einstein's equations, now known as the Kerr-Newman solution, describes spacetime around a rotating, electrically charged black hole. If you take the Kerr-Newman solution and set the charge equal to zero, you get Kerr's mathematical description of a rotating black hole; if instead you set the rotation equal to zero, you get the Reissner-Nordstrøm solution for a charged black hole; and if you set *both* charge and rotation to zero, you get Schwarzschild's solution for a non-rotating, uncharged black hole. The Kerr-Newman solution of Einstein's equations incorporates every property that a black hole can have – mass, charge and spin. In line with the no-hair theorem, it is the ultimate solution to those equations, at least as far as black holes are concerned. But since there is no reason to think that rotating black holes will actually have charge, any more than there is any reason to think that non-rotating black holes can have charge in the real Universe, I shall say no more about the Kerr-Newman solution, and will concentrate on the intriguing possibilities literally opened up by the addition of rotation alone to a massive black hole.

First, the ring singularity. Subject to the usual provisos about the mass of the black hole itself, and the size of the ring singularity (so that the astronaut is not torn apart by tidal forces) it would be possible to dive into a Kerr black hole at one of its poles, and right through the hoop formed by the singularity. As soon as you do so, the world is turned upside down. The equations tell us that as you pass through the ring you enter a region of spacetime in which the product of your distance from the centre of the ring and the force of gravity is *negative*. This might mean that gravity is behaving perfectly normally but you have entered a region of negative space, in which it is possible to be, for example, 'minus ten kilometres' away from the centre of the hole. Even relativists have trouble coming to terms with that possibility, so they usually interpret this negativity as meaning that *gravity* reverses as you pass through the ring, turning into a repulsive force that pushes on you instead of pulling. In the region of spacetime beyond the ring, the gravity of the black hole repels both matter and light away from itself, so that it acts like the white holes mentioned earlier.

This is uncomfortable enough to try to come to mental terms with; but there is another, even more uncomfortable implication in the equations describing this antigravity universe. An astronaut who dived through the ring but stayed close to it and circled around the centre of the black hole in an appropriate orbit would be travelling backwards in time. The saving grace, from the point of view of conventional physics, is that even if you did this, then dived back through the ring and on out of the rotating black hole, you still couldn't go back to the same region of spacetime that you

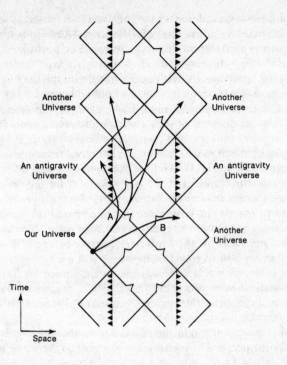

Figure 6.10 *The spacetime map of a rotating black hole is very similar to the one for a charged black hole, but includes an extra component – the antigravity universes. A. Allowed journeys. B. Forbidden journeys.*

started from. Like the event horizons of a Reissner-Nordstrøm black hole, those of a Kerr black hole allow only one-way travel, taking you on into another universe (Figure 6.10). You might arrive in that universe, in some sense, 'before' you left your original universe, but there would be no practical way of communicating with your starting place in order to pass a message to yourself before you left on the journey.

Nevertheless, just as you can imagine a charged black hole having so powerful an electric charge that the inner event horizon pushes out past the outer horizon and exposes the singularity inside, so a Kerr black hole that rotates sufficiently rapidly will fling off its event horizons and leave a naked singularity exposed to view. But this singularity, unlike that of a Reissner-Nordstrøm black hole, will still be in the form of a ring. It would be possible not only to travel through the ring, but to *look* through it from far away, using powerful telescopes. And if you did travel through the ring and into the region of negative time, there would no longer be any one-way horizons to prevent you going back again to your starting place. The Penrose diagram of such a situation is very simple. It consists of a negative universe and a positive universe, separated by a ring singularity through which anything can travel from one universe to the other (Figure 6.11). And, in principle, it would be possible to approach that singularity from any point of space and time in either universe, orbit around the singularity in the appropriate way, and return to exactly the same place that you started from, but getting back there before you had left. If one such naked Kerr singularity exists, anywhere in the Universe, then in principle it would be possible for you to travel from where you are sitting now to any place in the Universe and any time – past, present or future – that you wish, if only you could find the right route to follow. And, once again, none of this requires you to travel faster than the speed of light.

Of course, you might die of old age on the journey, but that is hardly the point. The equations of the general theory of relativity,

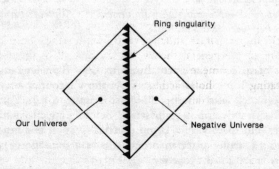

Figure 6.11 *A black hole that rotates fast enough will fling away its event horizon to expose a naked singularity connecting one 'negative' universe with one 'normal' universe.*

the best description of spacetime that we have, explicitly allow for the possibility of time travel. No wonder most physicists desperately wish that there really was an inviolable law of cosmic censorship – and that they are so concerned that there is no evidence at all that nature actually operates in accordance with such a law. But they can at least glean comfort from the fact that it would at the very least be extremely difficult to make a black hole spin fast enough for angular momentum to fling away its event horizons. Naked singularities of this kind may be impractical solutions to Einstein's equations, even if they are not, strictly speaking, impossible. Let's leave the bizarre properties of the ring singularity to one side, and look again at the overall spacetime map for a Kerr black hole.

Apart from this softening of the singularity, allowing a traveller to dive through the ring and back again, the spacetime map of the Kerr geometry is just like the paper doll topology of the Reissner-Nordstrøm geometry. Ignoring the region of negative time, we can represent this as in Figure 6.10, with the 'softness' of the singularity indicated by smoothing out the jagged 'shark's teeth' usually used to denote the world line of a singularity.

The bottom line of all these calculations, even leaving aside the puzzle of the antigravity, negative time regions of the map, is that objects which physicists are sure must exist in our Universe (rotating, massive black holes like the ones thought to be the powerhouses of quasars) provide connections through hyperspace to other regions of spacetime – other universes. How should we interpret these other universes? Is there really some infinite region of spacetime, in layer upon layer, going on 'forever' (whatever that means!)? In fact, it is equally valid, within the context of the equations of the general theory, to say that all of these different regions of spacetime are actually part of our own Universe, with the rotating black holes acting as hyperspace connections exactly like the wormholes developed from the notion of Einstein-Rosen bridges that I described on page 169. A rotating black hole may connect our Universe to itself, not just once but repeatedly, offering a gateway to different places *and different times*. The 'other universe' that you emerge into after a trip like the one shown in Figure 6.10 could actually be our own Universe, but a million years ago (or ten million years in the future). This is such a very alarming prospect, as far as our grasp on commonsense reality is concerned, that most physicists were mightily relieved when, in the 1970s, new calculations suggested that in the real Universe powerful gravitational effects associated with singularities and event horizons would choke off even these hyperspace connections before anything could pass through them. It seemed that wormholes would be able to exist only in an empty universe (in which case there would be nothing to pass through them, and no genuine possibility of using hyperspace connections for either space or time travel).

The blue-shift block

This problem with wormholes was first appreciated by mathematicians investigating the nature of white holes. One mathematician in particular, Douglas Eardley, of CalTech, seemed in the early 1970s to have proved conclusively that white holes cannot exist in the real Universe. This was particularly disappointing for me, because it pulled the rug from under a rather neat explanation of the way galaxies might have formed, a theory (developed by Soviet researchers in the 1960s) that I was particularly fond of, and even wrote a book about.*

The main exponent of the revival of the idea of white holes in the 1960s was Igor Novikov. He was intrigued by the evidence for explosive outbursts of activity in the Universe, such as the activity associated with quasars. At that time, nobody had fully worked out the way in which matter falling *into* a supermassive black hole could generate energy that would be channelled back *out* from the polar regions of the object,† and it seemed natural to some researchers to ask whether white holes might not be a better explanation for such phenomena than black holes. Novikov proposed that instead of the outburst from a singularity that created the big bang taking place all at once, there may have been pieces of the primordial singularity that somehow delayed their expansion, and burst out into the Universe at a late date. These 'lagging cores' would then spew matter out into the Universe in just the way we see quasars behaving. What's more, the gravity of a retarded core, even before it made its outburst, could have held on to a surrounding cloud of material in the expanding Universe; if stars formed in the cloud of gas around a lagging core, that could explain the origin of galaxies. This, alas, was the whole package of ideas that Eardley's work shot

---

* It was called, logically enough, *White Holes*.

† This is explained in *Cosmic Coincidences*, which I co-wrote with Martin Rees, the cosmologist who developed this model of quasar activity.

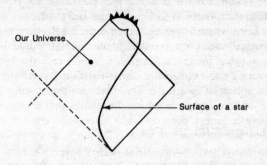

Figure 6.12 *The fly in the ointment. A black hole that forms in our Universe today does not have a past singularity, since the future singularity forms from the collapse of a star. This rules out all the interesting space- and time-travel possibilities, unless there are artificial means of opening the throat of the wormhole.*

down. We can begin to see why by looking at some more Penrose-type diagrams.

As well as black holes and white holes, relativists sometimes talk about 'grey' holes. A black hole is an object into which matter and radiation fall, but nothing escapes. A white hole is an object from which matter and radiation escape, but nothing falls in. A grey hole is an object from which matter and radiation escape, rise to a certain distance above the event horizon, and then fall back in.* In each case, remember that the black/white/grey hole is described by *two* singularities, one in the past and one in the future. The fact that this is really an idealized, mathematical way of looking at things is highlighted in Figure 6.12, which shows the relevant spacetime diagram representing the collapse of a real massive star to form a black hole. Spacetime is only really described accurately by the Schwarzschild solution to Einstein's equations in the region above (or outside) the surface of the star. The star itself cuts off a large part of the spacetime diagram, to the right in Figure 6.12, from having any real significance. It is only when the star collapses that the Schwarzschild metric really comes into its own, and the future

* This is rather like the description of the Universe as emerging from a big bang, expanding for a while, and then contracting back into a big crunch. We may live in a grey hole!

singularity alone has the possibility of real existence. For a realistic collapsing star, there is no past singularity and no past event horizon from which anything can emerge. Of the three permissible mathematical variations on the theme, only the black hole is a realistic physical prospect.

Of course, if the collapsing star is rotating fast enough, we still have the option of creating a Kerr black hole providing a gateway to some other universe in which the matter that collapses into the black hole in our Universe could re-emerge from a past event horizon in that universe as a white hole; but there are problems with this scenario as well. One concerns Hawking radiation. Singularities that lie horizontally across a spacetime diagram in the future (such horizontal singularities are called 'spacelike', because they exist across all of space but only at one moment in time) do not suffer the consequences of Hawking evaporation. From the perspective of such a singularity, all time lies in the past, and there is no future into which Hawking evaporation can take place (always assuming that the flow of time cannot be reversed, which is still the subject of some speculation). A spacelike singularity in the past, on the other hand, can produce a plethora of particles by the Hawking process, maybe even evaporating itself away to nothing at all. The fate of these particles, of course, is to fill up the inside of the black hole, and to fall inevitably together to form a spacelike future singularity. This doesn't really alter the picture of a Schwarzschild black hole very much, although it does shed some new light on what may be going on inside the hole, where (it used to be thought) nothing interesting happened at all. The real snag comes when we apply the same reasoning to the kind of 'vertical' (or 'timelike') singularity associated with rotating or charged black holes. It is, after all, the fact that the future singularity is turned on its side to become a timelike singularity that makes it possible, in principle, to navigate a spacecraft into a black hole and out again into another universe, without being crushed by gravity. But if the timelike singularity itself evaporates by the Hawking process, what happens to all the particles produced? Once again, according to what some physicists argue is the simplest interpretation of the equations, they must fill up the inside of the black hole and pile up at a moment in the future, forming a spacelike future singularity and barring the way to other universes.

I must say that I am not entirely convinced by these arguments. It is a key feature of Hawking evaporation, in its original form, that it involves processes going on at the event horizon, so that when pairs of particles are produced one partner can escape while the other drops inside the hole into a negative energy state. It is not at all obvious that the same sort of process will be at work beneath the event horizon, next door to what amounts to a naked singularity. Still, more eminent mathematicians than I am seem to take the notion seriously, and if they are right to do so then it seems that quantum processes may close the door to other worlds opened by general relativity. But since we do not yet have a complete theory combining quantum physics and general relativity in one set of equations, this disappointing conclusion cannot yet be regarded as the last word on the subject. We can see just how dramatically conclusions about black holes can be turned on their heads by looking at what happened to Eardley's work, which seemed, at the time he presented it, to put the lid on the possibility of white holes even on the basis of general relativity alone.

The important point highlighted by his more realistic look at the collapse of a star to form a black hole is that we have to take account of the real distribution of matter in the Universe outside, not just of the elegant equations describing curved spacetime. Problems of this kind do not arise when we are describing the big bang itself, because there was no outside, and therefore no outside matter or energy, to worry about. But things are different for a retarded core. I have already mentioned that one of the attractive features of Novikov's idea was that the gravity of the retarded core would hold on to matter, perhaps explaining the presence of galaxies in the expanding Universe. The snag is, such a core would hold on to matter, and even light, *too* effectively. Light escaping from the surface of a black hole, remember, is red-shifted so much that it loses all its energy. The red shift is infinite. But light falling on to a black hole gains energy, and as it crosses the event horizon it is infinitely *blue*-shifted. This is of no concern to us as long as the build-up of energy is safely locked away inside a black hole (although it does have intriguing cosmological implications, which I shall discuss in Chapter Eight). But now think about what happens to a white hole, in a real universe already containing matter and energy, as it tries to emerge from a singularity.

The expanding core of the white hole will possess a gravitational field every bit as powerful as the equivalent black hole. So matter and energy will come piling in on top of the object from the Universe outside, even as the white hole inside tries to expand outwards. The problem is particularly acute for any retarded cores left over from the big bang, since in the fireball of creation they would have been surrounded by a seething maelstrom of energy which they could feed off; but Eardley showed that even in the Universe today there is ample energy available, in the form of starlight alone, to provide a pile-up at the event horizon. After all, if the blue shift is *infinite*, you only need a tiny amount of light falling into the white hole for the blue shift to cause problems. Those problems take the form of what is now known as a 'blue sheet', a wall of energy surrounding the white hole, so intense that the energy of the light itself warps spacetime so much that it creates a black hole surrounding the incipient white hole. As Stanford physicist Nick Herbert has graphically expressed it, 'universes like our own contain lethal amounts of light and matter which will form fatal blue sheets that smother infant white holes in their cradles.' More prosaically, the calculations suggest that the smothering process would take about one thousandth of a second if any lagging core in the Universe today decided to stop lagging and tried to become a white hole.

Worse, the smothering carries over from the Schwarzschild solution to the Reissner–Nordstrøm and Kerr solutions. Such holes always, of course, have past event horizons. The pile-up of energy at the past event horizon begins at the moment the Universe (and the horizon) is created, and forms an impenetrable blue sheet. Nobody has yet completely solved the difficult mathematical problem of describing accurately how this blue sheet interacts with the wormhole, but at the end of the 1980s most physicists regarded the existence of such a blue sheet as likely to cut the connection between universes. Imagine their surprise, then, when calculations carried out at the very end of that decade, and the beginning of the 1990s, suggested that this might not, after all, be the case.

Parting the blue sheet

This work was carried out only after the investigation of traversable wormholes by Thorne and his colleagues which had been prompted by Sagan's enquiry. But it follows on logically from Eardley's work on blue sheets, so it makes sense to discuss it first, before (I promise!) getting back to the science fictional scenario and the factual breakthroughs stemming from it.

Eardley showed that problems with blue sheets arise in the real Universe, because as well as considering the curvature of spacetime around a singularity, we have to allow for the way that curved spacetime interacts with matter and energy from the Universe outside. But how does it interact with the *spacetime* outside? Those calculations assumed that the spacetime outside the black/white hole is itself flat. This is so nearly the case, for regions of space on the scale of our Solar System or the Milky Way Galaxy, that it is almost taken for granted by the experts – it is certainly the picture they use as a first approximation to reality. But it may *not* be the case on the scale of the Universe itself. Einstein's cosmological equations, which tell us the Universe must be either expanding or contracting, also tell us that the geometry of the Universe is highly unlikely to be flat. It is much more likely to be non-Euclidean and curved – either open, like the saddle surface discussed in Chapter Two, or closed, like the surface of a sphere. And researchers at the University of Newcastle upon Tyne have shown that if indeed the Universe is closed (the option favoured by most cosmologists today, for reasons detailed in *Cosmic Coincidences* and touched on in Chapter Eight), then there may, after all, be holes in the blue sheet story, if not in the blue sheets themselves.

Journey into hyperspace

Because the Reissner-Nordstrøm solution is easier to work with than the Kerr solution, these investigations have concentrated, so far, on the behaviour of charged black holes in a realistic mathematical model of the Universe. It is expected that the important properties associated with the existence of two event horizons around each such black hole will carry over into the Kerr solution for rotating black holes, but these are still early days in this work and there could conceivably be further surprises when rotation is added in to the calculations. The problems involving blue sheets arise, in the old picture (where 'old' now means pre-1988), at the inner event horizon, which is also known as the Cauchy horizon. They can be explained, in physical terms, as a result of an observer sitting at the Cauchy horizon and seeing the entire infinite future of the outside universe in a finite time on the observer's clocks. But suppose the outside universe does not have an infinite future! What if it is finite and unbounded, like the closed surface of a sphere?

This possibility was initially investigated chiefly by Felicity Mellor, at Newcastle, working with Ian Moss, a former protégé of Stephen Hawking, and Paul Davies, who was then Professor of Physics at Newcastle but is now based in Adelaide, in Australia. They looked at the mathematical description of wormholes associated with charged black holes in the geometry corresponding to a closed universe – one which has its own cosmological event horizon. In other words, they had to deal with *three* event horizons, two associated with the black hole and one cosmological. The particular cosmological models they studied also include another feature, related to the constant that Einstein tried to use to fiddle his equations to make the model universes of the general theory hold still. But this modern version of the cosmological constant, far from holding the Universe still, is invoked to explain how it expanded away from the intense gravity of the initial singularity. It operates way back at the edge of universal time, close to the singularity in which the Universe was born, and acts as a kind of negative-pressure antigravity, whooshing the embryonic Universe up from a volume far smaller than that of an atom to about the size of a grapefruit in a tiny fraction of a second, before fading away as the Universe settles down into the more steady expansion we see today. The phase of very rapid expansion is known as inflation, and is a key ingredient of the modern version of big bang theory. All the Newcastle team's conclusions about wormholes hold up, Moss showed, provided only that the Universe is closed; but it would have been a deep embarrassment if they did not work in the context of the inflationary scenario, which is the current 'best buy' in cosmology, so it would have been foolish not to check the calculations with the presence of this kind of cosmological constant. In such scenarios, space away from concentrations of matter is very nearly flat, and is called de Sitter space; but spacetime itself can still be gently curved around to make a closed Universe. The spacetime is like two black holes at opposite 'ends' of the Universe. Mellor and Moss found that in these circumstances the Universe can contain many black holes separated by regions which correspond almost exactly to de Sitter space, and that these black holes can (if charged) be connected by wormholes which are stable. In some cases, naked singularities can form, violating cosmic censorship; and, in the Newcastle team's own words, 'an observer could in principle travel through the black hole to another universe.'

The main contribution Paul Davies made to this work was to

include an allowance for quantum effects. As Hawking demonstrated so vividly in the 1970s, quantum effects can have a dramatic impact on the behaviour of black holes, and it was natural to wonder whether they would prevent the kind of wormholes described by Mellor and Moss from making an appearance in the real Universe. But no. In 1989 Davies and Moss reported that 'the conjecture that an object may pass through a black hole and enter "another universe"' still holds for charged black holes in a closed Universe even when quantum effects are taken into consideration. As long as the Universe is closed, neither the presence of a cosmological constant nor the quantum complications prevent traversable wormholes from existing, and 'the Mellor-Moss solutions might provide genuine "space bridges" to other universes.'★

All of this work concerns natural features of the Universe – black holes formed naturally, like those associated with quasars, or left over from the superdense state of the big bang itself. If all of the maths stands up when somebody carries out the difficult task of adapting the calculations to cover rotating black holes, it will mean that hyperspace connections can arise naturally in a universe like ours. And this astonishing discovery provides a dramatic backdrop to the speculations, encouraged by Sagan's wishful thinking and developed by the CalTech researchers and others, that it might indeed be possible to construct traversable wormholes *artificially*, just as Sf writers have been telling us for decades, given a suitably advanced technological civilization.

Wormhole engineering

There is still one problem with wormholes for any hyperspace engineers to take careful account of. The simplest calculations suggest that whatever may be going on in the universe outside, the actual passage of a spaceship through the hole (or, rather, the *attempted* passage of a spaceship through the hole) ought to make the star gate slam shut. The problem is that, even leaving aside the question of radio waves or light from the spaceship piling down on to the singularity and creating an infinite blue sheet, an accelerating object, according to the general theory of relativity, generates those ripples in the fabric of spacetime itself known as gravitational waves. It is the effect of this gravitational radiation, pouring out

★ *Classical and Quantum Gravity*, Volume 6, pp. L173–L177.

into space from the binary pulsar, that is draining away energy so that the orbit of the pulsar changes measurably, providing the best confirmation yet of the accuracy of Einstein's theory. Gravitational radiation itself, travelling ahead of the spaceship and into the black hole at the speed of light, could be amplified to infinite energy as it approaches the singularity, warping spacetime around itself and shutting the door on the advancing spaceship. Even if a natural traversable wormhole exists, it seems to be unstable to the slightest perturbation, including the disturbance caused by any attempt to pass through it.

But Thorne's team found an answer to that for Sagan. After all, the wormholes in *Contact* are definitely not natural; they are engineered. One of his characters explains:

> There is an interior tunnel in the exact Kerr solution of the Einstein Field Equations, but it's unstable. The slightest perturbation would seal it off and convert the tunnel into a physical singularity through which nothing can pass. I have tried to imagine a superior civilization that would control the internal structure of a collapsing star to keep the interior tunnel stable. This is very difficult. The civilization would have to monitor and stabilize the tunnel forever.*

But the point is that the trick, although it may be very difficult, is not impossible. It could operate by a process known as negative feedback, in which any disturbance in the spacetime structure of the wormhole creates another disturbance which cancels out the first disturbance. This is the opposite of the familiar positive feedback effect, which leads to a howl from loudspeakers if a microphone that is plugged into those speakers through an amplifier is placed in front of them. In that case, the noise from the speakers goes into the microphone, gets amplified, comes out of the speakers louder than it was before, gets amplified . . . and so on. Imagine, instead, that the noise coming out of the speakers and into the microphone is analysed by a computer that then produces a sound wave with exactly the opposite characteristics from a second speaker. The two waves would cancel out, producing total silence. For simple sound waves, pure notes that correspond to waves like the sine curve in Figure 5.8 (p. 158), this trick can actually be carried out, here on Earth, in the 1990s. Cancelling out more complex noise, like the

* Legend edition, p. 347.

roar of a football crowd, is not yet possible, but might very well be in a few years' time. So it may not be completely far-fetched to imagine Sagan's 'superior civilization' building a gravitational wave receiver/transmitter system that sits in the throat of a wormhole and can record the disturbances caused by the passage of the spaceship through the wormhole, 'playing back' a set of gravitational waves that will exactly cancel out the disturbance, before it can destroy the tunnel.

But where do the wormholes come from in the first place? The way Morris, Yurtsever and Thorne set about the problem posed by Sagan was the opposite of the way everyone before them had thought about black holes. Instead of considering some sort of known object in the Universe, like a dead massive star, or a quasar, and trying to work out what would happen to it, they *started out* by constructing the mathematical description of a geometry that described a traversable wormhole, and *then* used the equations of the general theory to work out what kinds of matter and energy would be associated with such a spacetime. What they found is almost (with hindsight) commonsense. Gravity, an attractive force pulling matter together, tends to create singularities and to pinch off the throat of a wormhole. The equations said that in order for an artificial wormhole to be held open, its throat must be threaded by some form of matter, or some form of field, that exerts negative pressure, and has antigravity associated with it.

This already has echoes of the kind of field, associated with the modern version of a cosmological constant, thought to have driven the expansion of the very early Universe; I shall return to this intriguing connection shortly. The critical factor for keeping a wormhole open is that the negative pressure (or tension) exerted must be greater than the mass-energy density of the original matter that makes up the black hole. In other words, the antigravity associated with the negative pressure more than cancels out the effects of gravity, inside the wormhole itself. For a hole a few kilometres across (roughly the size of a neutron star), the negative pressure must be stronger than the normal pressure at the heart of a neutron star. Hardly surprisingly, hypothetical matter that possesses this curious property is known as 'exotic' matter. The CalTech team showed that any traversable wormhole must contain some form of exotic stuff. The work by Mellor, Moss and Davies *may* weaken this restriction, since their investigations suggest that natural wormholes can exist even without the aid of exotic matter.

But since we are interested in artificial wormholes (a supercivilization could not rely on finding natural hyperspace connections in just the places it needed them, and in any case there are other obvious difficulties about approaching the centres of quasars), there really seems no escape from the need for exotic matter.

Now, you might think, remembering your high-school physics, that this completely rules out the possibility of constructing traversable wormholes. Negative pressure is not something we encounter in everyday life (imagine blowing negative-pressure stuff *into* a balloon and seeing the balloon *deflate* as a result). Surely exotic matter cannot exist in the real Universe? But you may be wrong. The evaporation of black holes in the Hawking process actually involves negative energy states, remember, and this is equivalent to a kind of negative pressure operating at the horizon of a black hole; and there is another way in which negative pressure not only *can* be produced in theory but *has* been produced, and measured, in the laboratory.

Making antigravity

The key to antigravity was found by a Dutch physicist, Hendrik Casimir, as long ago as 1948. Casimir, who was born in The Hague in 1909, is best-known for his work on superconductivity, a strange phenomenon in which some materials, when cooled to very low temperatures, lose all their electrical resistance (physicists and engineers have recently been excited by the discovery that some superconductors do not have to be supercold but can operate at relatively high temperatures, although not yet quite at ordinary room temperature). From 1942 onwards, Casimir worked in the research laboratories of the electrical giant Philips, and it was while working there that he suggested an even stranger possibility than superconductivity, implicit in the rules of quantum physics, that became known as the Casimir effect.

The simplest way to understand the Casimir effect is in terms of two parallel metal plates, placed very close together with nothing in between them (Figure 6.13). But as we have already seen, the quantum vacuum is not like the kind of 'nothing' physicists imagined the vacuum to be before the quantum era. It seethes with activity, with particle-antiparticle pairs constantly being produced

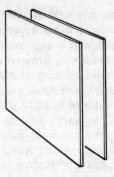

Figure 6.13 *The physics that may make it possible to hold wormholes open can be seen at work when two simple metal plates are placed close to each other in a vacuum.*

and annihilating one another. Among the particles popping in and out of existence in the quantum vacuum there will be many photons, the particles which carry the electromagnetic force, some of which are the particles of light. Indeed, it is particularly easy for the vacuum to produce virtual photons, partly because a photon is its own antiparticle, and partly because photons have no 'rest mass' to worry about, so all the energy that has to be borrowed from quantum uncertainty is the energy of the wave associated with the particular photon. Photons with different energies are associated with electromagnetic waves of different wavelengths, with shorter wavelengths corresponding to greater energy; so another way to think of this electromagnetic aspect of the quantum vacuum is that empty space is filled with an ephemeral sea of electromagnetic waves, with all wavelengths represented.

This irreducible vacuum activity gives the vacuum an energy, but this energy is the same everywhere, and so it cannot be detected or used. Energy can be used to do work, and thereby make its presence known, only if there is a *difference* in energy from one place to another. A good example is the way electricity is used to light your home. In the lighting circuit, one wire is kept at a modestly high electrical potential energy (perhaps corresponding to 110 volts, or 240 volts, depending on where you live), while another (the 'earth') is at the zero of electric energy. The energy inherent in the higher voltage wire does nothing at all until a connection is made to the low-voltage wire – this is why it is known as 'potential' energy. When a connection is made, electricity flows across the connection, releasing potential energy as actual energy in the form of heat and light. The potential *difference* is crucially important, and if both wires are at the same voltage, whether it is zero or 240 volts, or even larger, no current will flow. Indeed, if the whole world were charged up to a couple of hundred volts we wouldn't all glow with electrical energy, because there would be no lower-energy place for the electricity to drain into. Such an electrically charged planet would resemble, in a crude sense, the way in which the vacuum is uniformly packed with energy. That is called, logically enough, the vacuum energy; and Casimir showed how to make it visible.

Between two electrically conducting plates, Casimir pointed out, electromagnetic waves would only be able to form certain stable patterns. Waves bouncing around between the two plates would behave like the waves on a plucked guitar string. Such a string can only vibrate in certain ways, to make certain notes – ones for which the vibrations of the string fit the length of the string in such a way that there are no vibrations at the fixed ends of the string. The allowed vibrations are the fundamental note for a particular length of string, and its harmonics, or overtones. In the same way, only certain wavelengths of radiation can fit into the gap between the two plates of a Casimir experiment (Figure 6.14). In particular, no photon corresponding to a wavelength greater than the separation between the plates can fit into the gap. This means that some of the activity of the vacuum is *suppressed* in the gap between the plates, while the usual activity goes on outside. The result is that in each cubic centimetre of space there are fewer virtual photons bouncing around between the plates than there are outside, and so the plates feel a force pushing them together. Unfortunately, because the excluded photons are the ones with longer wavelengths, and therefore lower energy, the effect is very small. But the force does exist, and shows itself as a force of attraction between the two plates, sucking them together – negative pressure.

It may sound bizarre, but it is real. Several experiments have been carried out to measure the strength of the Casimir force between two plates, using both flat and curved plates made of various kinds of material. The force has been measured for a range of plate gaps from 1.4 nanometers to 15 nanometers (one nanometer is one billionth of a metre) and exactly matches Casimir's prediction.

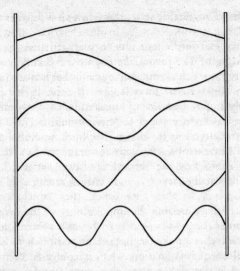

Figure 6.14 Only certain lengths of wave can fit into the gap between the metal plates represented in Figure 6.13.

Another scientist who, like Sagan, writes science fiction is Robert Forward, of the Hughes Research Laboratories in Malibu, California. He has suggested that the Casimir effect might even be put to practical use, extracting energy from the vacuum. Unlike Sagan, Forward is possibly even better known as a science fiction writer than as a scientist. A larger-than-life character, he is the kind of scientist who speculates about ways to use antimatter in the propulsion systems of spacecraft, and the kind of Sf writer who describes life forms that have evolved on the surface of a neutron star. To him, extracting energy from the vacuum – out of what we used to think of as nothing at all – is easy.

Forward's design for a 'vacuum-fluctuation battery' consists of a helix of ultrathin aluminium foil charged up with electricity; it would look rather like the toy known as a slinky. The positive electric charge holds the leaves of foil apart, while the Casimir force tries to pull them together. In this state, if the whole thing is allowed to collapse slowly like an accordion being squeezed, energy from the Casimir force will be released as usable electricity. Once the 'accordion' has collapsed, the 'battery' can be recharged using electricity from an outside source, just like an ordinary rechargeable battery.

Of course, Forward's vacuum-fluctuation battery (described in the sober pages of the journal *Physical Review B* in August 1984*) is really totally impractical; but, once again, that is not the point. It is allowed by the laws of physics, and it depends entirely on the proven reality, albeit on a very small scale, of the phenomenon of negative pressure. In a paper they published in 1987, Morris and Thorne drew attention to such possibilities, and also pointed out that even a straightforward electric or magnetic field threading the wormhole 'is right on the borderline of being exotic; if its tension were infinitesimally larger . . . it would satisfy our wormhole-building needs.'† In the same paper, they concluded that 'one should not blithely assume the impossibility of the exotic material that is required for the throat of a traversable wormhole.' The two CalTech researchers make the important point that most physicists suffer a failure of imagination when it comes to considering the equations that describe matter and energy under conditions far more extreme than those we encounter here on Earth. They highlight this by the example of a course for beginners in general relativity, taught at CalTech in the autumn of 1985, after the first phase of work stimulated by Sagan's enquiry, but before any of this was common knowledge, even among relativists. The students involved were not taught anything specific about wormholes, but they were taught to explore the physical meaning of spacetime metrics. In their exam, they were set a question which led them, step by step, through the mathematical description of the metric corresponding to a wormhole. 'It was startling', said Morris and Thorne, 'to see how hidebound were the students' imaginations. Most could decipher detailed properties of the metric, but very few actually recognized that it represents a traversable wormhole connecting two different universes.'

For those with less hidebound imaginations, there are two remaining problems – to find a way to make a wormhole large enough for people (and spaceships) to travel through, and to keep the exotic matter out of contact with any such spacefarers. The best suggestion along these lines, raising the real possibility that superior civilizations might indeed be able to manufacture their own hyper-

* Volume 30, pp. 1700–1702.
† *American Journal of Physics*, Volume 56, pp. 395–412.

space connections, has come from Matt Visser, of Washington University in St Louis, Missouri. The key ingredient is string.

The string-driven spaceship: a practicable proposition?

If current ideas about the birth of the cosmos are correct, the expanding Universe as we see it is a more leisurely product of the phase of wild expansion, driven by negative-pressure antigravity in the form of a cosmological constant, which occurred during the first split-second after the moment of creation. The expansion slowed to its more sedate rate, as we see it today, when the fields associated with that cosmological constant decayed into other forms and disappeared, taking the constant with them. But there is no reason to think that this transition from the inflationary epoch to the present state of expansion would have occurred perfectly uniformly and smoothly, at the same instant everywhere across the embryonic Universe. Quite the reverse. Cosmologists calculate that the changes in the fields associated with this transition are more likely to have occurred independently in many different, distinct regions of the young Universe, known as domains. Within each domain, the changeover would have been quite smooth. But at the boundaries between domains, the leftover fields from the decay of the negative-gravity fields would not have fitted together smoothly, causing distortions in the structure of spacetime.

These distortions are rather like the features known as dislocations which often occur in crystals. The distinctive thing about a perfect crystal is that the atoms it is made up of are lined up in orderly rows. When a crystalline solid forms from a cooling liquid, however, the liquid does not all solidify into one perfect crystal. Instead, different regions crystallize out in slightly different ways, just like the cosmologists' domains, so that the atoms of the solid are arranged in slightly different orientations in different places. The orderly rows in one domain do not match up with those in the domain next door, and there is a kind of fault line where the two rows of atoms meet. The boundaries between different regions can appear as flat planes, along which it is easy to cleave the crystal in two, or as hairlike lines.

According to calculations of the physics of the very early Universe, the same sort of thing would have happened at the cosmological transition, which is sometimes described as the equivalent of the freezing of a liquid into a solid – the difference being that the 'liquid' and 'solid' involved in the cosmological transition correspond to different states of the vacuum itself. As different domains 'froze out' slightly differently, spacetime defects in the form of walls across the Universe, thin tubes, and even mathematical points should have formed. We don't see any walls of this kind in our part of the Universe, the story runs, because the Universe has expanded so much since that time that they are out of sight. And we don't see the point defects, because they are hard to find (although some versions of the theory suggest that they could show up as isolated magnetic poles – a north without a south, or vice versa). But the tubes left over from this epoch may very well still exist in our part of the Universe, and may even play an important role in determining the distribution of matter in the Universe as we see it.

Such tubes are known as cosmic strings. They cannot have ends, but must always either form closed loops or stretch right across the observable Universe. Such strings, if they exist, are very thin indeed – one thousand-billion-billion-billionth of a centimetre across. And yet, a piece of cosmic string just a kilometre long would weigh as much as the Earth. A string that stretched right across the Universe and was ten billion light years long could be scrunched up into a ball smaller than a single atom, but it would weigh as much as a supercluster of galaxies. Some astronomers suggest that the presence of loops of string, early in the life of the Universe, could have provided the gravitational 'seeds' on which galaxies and clusters of galaxies grew. Because of their strong gravitational pulls, the string loops would hold back material from the cosmic expansion, allowing it to form stars and galaxies. But this strong gravitational pull is only an outward feature of the string, and, in the context of building a traversable wormhole, its least interesting property. What is far more interesting is what lies *inside* the tube.

The simplest way to think of the inside of a cosmic string is that it is a thin tube containing material left over from the earliest phase of universal expansion, before the transition into the present state. Cosmic strings are full not of matter, but of the original energy fields themselves, like fossils from the first split-second. And those fields still carry the stamp of the cosmological constant, the enormous negative pressure that extended everywhere when the

Universe was born. In a stretched piece of elastic, the tension in the material tries to pull the ends together. In a stretched cosmic string, the negative tension associated with the negative pressure tries to stretch the string even more. The inside of a cosmic string is exotic stuff with all the power of anything you could possibly need to stabilize a wormhole.

Matt Visser's leap of the imagination was to dispense with the assumption of spherical symmetry, which the relativists usually use to make their calculations easier. In an essay for the Gravity Research Foundation Competition in 1989 (Visser's essay didn't win, but received an Honorable Mention), he took a leaf from the book of Thorne's team, and designed a spacetime structure that would allow for easy passage through the wormhole, *then* worked out how to place the exotic stuff to produce such a structure. Because we are dealing with two three-dimensional spaces (two universes or two parts of the same universe) connected by the star gate, the surfaces that form the entrances and exits to the wormholes have to be three-dimensional. Previously, I have described these in terms of spherical black holes, perhaps with the addition of rotation, forcing the spherical surfaces to bulge a bit at the equator. But Visser, an expert theoretical hyperspace engineer, decided he would like a flat surface for his travellers to pass through, with no strong gravitational fields to disturb them, and with the exotic stuff kept well out of their way. The structure he came up with is the six-sided surface of a cube, with all the exotic stuff confined to struts along the edges of the cube. A traveller who approaches and crosses one face of such a cube will feel no tidal forces, and will encounter no matter, exotic or otherwise. 'Such a traveller', says Visser, 'will simply be shunted across the universe' and will emerge from an equivalent cube in another region of flat space – perhaps even in another universe.

Cosmic strings are not specifically mentioned in that essay, nor in the more formal mathematical version of these ideas that Visser published in the *Physical Review D*.* But in that formal paper he does point out that 'the stress-energy present at the edges of the cube is identical to the stress-energy . . . of a *negative tension* classical string' (his italics). 'No natural mechanism for generating negative string tension [today] is currently known', says Visser; but there is, of course, a known mechanism which may have generated

* Volume 39, pp. 3182–4, 1989.

negative string tension long ago in the birth of the Universe. Where better to get the exotic stuff to lay out along the struts of his cubical entrances to traversable wormholes?

Any prospect of building such a device is far beyond our present capabilities. But, as Morris and Thorne stress, it is not impossible and 'we correspondingly cannot now rule out traversable wormholes.' It seems to me that there's an analogy here that sets the work of such dreamers as Thorne and Visser in a context that is both helpful and intriguing. Almost exactly 500 years ago, Leonardo da Vinci speculated about the possibility of flying machines. He designed both helicopters and aircraft with wings, and modern aeronautical engineers say that aircraft built to his designs probably could have flown if Leonardo had had modern engines with which to power them – even though there was no way in which any engineer of his time could have constructed a powered flying machine capable of carrying a human up into the air. Leonardo could not even dream about the possibilities of jet engines and routine passenger flights at supersonic speeds. Yet Concorde and the jumbo jets operate on the same basic physical principles as the flying machines he designed. In just half a millennium, all his wildest dreams have not only come true, but been surpassed. It might take even more than half a millennium for Matt Visser's design for a traversable wormhole to leave the drawing board; but the laws of physics say that it is possible – and, as Sagan speculates, something like it may already have been done by a civilization more advanced than our own.

There are still, of course, practical difficulties involved. Even if Sagan's superior civilization had the capabilities required to manipulate cosmic string, and knew where to find it, there would still be the little problem of travelling across space to wherever the string was located in order to get the stuff to build the star gate. If you can travel far across space in any case, perhaps the star gate would be unnecessary; if you can't travel far across space by other means, perhaps you could never get hold of the raw materials to build your star gate. But even if you already have some other efficient means of space travel, there may be another powerful incentive to try to build a traversable wormhole. In a 'note added in proof' near the end of their *American Journal of Physics* paper about using wormholes for interstellar travel, Morris and Thorne commented that 'since writing this, we have discovered that from a single wormhole an arbitrarily advanced civilization can construct a machine for backward time travel.' In other words, every star gate is also a potential time machine. Incredibly, though, this is only half the story of time travel. For there is another, quite separate way in which the laws of physics allow for the possibility of travel backwards in time, an idea developed in detail in a paper published in the *Physical Review D* (Volume 9, p. 2203) fully *fifteen years* before Morris and Thorne added that note in proof to their first epic wormhole paper. The general theory of relativity tells us, in fact, that there are two ways to build a time machine – so let's look at them both in detail.

CHAPTER SEVEN

Two Ways to Build a Time Machine

How commonsense doesn't make sense. The granny paradox – and how to doctor it. Schrödinger's cats and the many-worlds theory. Tangling time. Is time an illusion? Time-travelling tachyons. A universal time machine, Tipler's time machine, and time tunnels, Soviet-American style. Spacetime billiards and cosmic histories – adding two plus two (plus many more) Richard Feynman's way

Commonsense tells us that time travel is impossible. Commonsense also tells us that it is nonsense to suggest that moving objects shrink and get heavier, and that an astronaut who travels to a distant star and returns to Earth will be younger than her twin brother who stayed at home. Commonsense is *not* always a good guide to the laws on which the Universe operates, and when it comes to time travel, as with anything else, it is important to find out what those laws really tell us, not what we would like them to say. But that doesn't mean that we can completely dismiss the doubts about time travel expressed by philosophers and implicit in our commonsense view of the notion. If time travel *is* possible, it will certainly mean abandoning some cherished beliefs about the nature of reality – but it won't be the first time that physicists have had to do that in the past hundred years.

By 'time travel', of course, I mean *two-way* travel in time, some process that will enable you to go on a journey and return to the same place you started from at the moment you left (or before).

Such a time journey is said to form a closed timelike loop, or CTL. In 'commonsense' terms, the problem with this kind of time travel is graphically illustrated by imagining what would happen to a time traveller if he or she travelled back in time and somehow contrived (or inadvertently caused) the death of their own maternal grand-mother, before the time traveller's mother had been born. In that case, the time traveller could never have been born. So the journey could never have been made, and granny has not died after all. In which case, the time traveller *has* been born . . . and so on.

Paradoxes and possibilities

In more scientific terminology, the problem of closed timelike loops is that they may violate causality. Causality is a hypothetical law which says that causes always precede their effects. If I flip the switch on the wall by the door to my room, the light comes on after I flip the switch, not before. Even within the conventional frame-work of relativity theory, which allows observers moving at different speeds to see the same events (in some cases) occurring in different sequences or at different times, no observer, however he or she is moving, will ever see the light in my room come on just before I flip the switch. Think of a moving railway carriage with a light source in the middle. Different observers may disagree about whether the two pulses of light from the source reach the two ends of the carriage at the same time, or which pulse gets to its appropriate end first; but all observers agree that the pulses leave the light source before they arrive at the end walls. Most physicists believe that causality is an inviolable law of nature; but, in fact, they have no proof that this is the case. Nobody has ever seen causality being violated, but, as with the cosmic censorship 'rule', there is actually nothing in the laws of physics that requires causality to be true. The causality law is no more than our commonsense view of time expressed in scientific jargon.

So how might we resolve the 'granny paradox'? There are two well-established possibilities, which have been widely discussed by scientists, philosophers and (most accessibly) by science fiction writers. The first is that the past may be inviolable, already set in a rigid pattern. Everything that has happened, including your voyage back in time to visit granny, has, on this view, already happened and cannot be altered. So, whatever your intentions when you set off on your journey, nothing you do will change the past. Should you set out with murderous intent, it may be that your gun will misfire when you take aim at granny; or perhaps, through a series of seemingly chance events, you will never actually get to meet her at all.

A slight variation on this idea is that it may be possible to go back in time and change the past, but not in any significant sense. For example, if you were to go back and chop down a tree, another would grow in its place; if you murder granny as a young girl, your grandfather might marry her sister instead, so that there is only a minor change in the genetic material you inherit; and so on. Fritz Leiber, in his 'Change War' series of stories, has two opposing groups of time travellers struggling to defeat each other by changing the past, each to their own advantage. Try as they may, though, the changes they make seem to have little influence, and 'damp out' before they spread very far through the spacetime continuum – in obedience to what one of Leiber's characters refers to as 'the Law of Conservation of Reality'.* The most worrying aspect of this resolution of the granny paradox is the extent to which it seems to remove our capacity for free will and truly independent action. If the past is so rigidly fixed in place, along with all CTL trips, maybe the future is equally fixed as well, and our perception of time flowing, with decisions we make affecting the outcome of events, is no more real than the appearance of lifelike motion and a flow of time generated when the still pictures that make up a film are projected on to a screen in rapid sequence.

This idea of time as in some sense a fixed and unalterable dimension seems to have been first propounded by H. G. Wells in his famous story *The Time Machine*, which first appeared in book form in 1895. Exactly ten years before Einstein published his special theory of relativity, and even longer before Minkowski described the special theory in terms of four-dimensional spacetime geometry, Wells wrote that 'there is no difference between Time and any of the three dimensions of Space, except that our consciousness moves along it.' The fictional time traveller of the story describes what we perceive as a three-dimensional cube as in fact being a fixed and unalterable four-dimensional entity, extending through time and therefore having as its dimensions length, breadth,

* 'Try and Change the Past', in *Trips in Time*, edited by Robert Silverberg (Nelson, New York, 1977); see also *The Big Time*, Fritz Leiber (Ace Books, New York, 1961).

thickness *and duration*. But the problem with all this is, if everything is fixed in four dimensions, how can the traveller have any influence on the events he becomes involved with later in the story? According to Wells' own justification for the adventures, everything, including the traveller's intervention in the future, is already fixed and predetermined. Which seems to take most of the fun out of life.

The second possibility for resolving the granny paradox is more intriguing. It is now well established that at the subatomic level the Universe is governed by quantum rules which operate in accordance with the laws of chance and probability. Again, there is a hackneyed (but powerful) way to understand what this means. The decay of the nucleus of a radioactive atom, with the nucleus spitting out a particle and becoming the nucleus of an atom of a different element, is governed entirely by chance. For each particular type of radioactive element, there is a specific length of time during which there is a precise 50:50 chance that the atom will decay. This time interval is known as the half-life of the element. The slavish obedience of such quantum process to the laws of probability deeply affronted Einstein, and led to his famous comment 'I cannot believe that God plays dice with the Universe'; but all the evidence (and there is a great deal of it) is that probability does indeed rule at the quantum level. The classic thought experiment which brings home the bizarre implications of this was dreamed up by Nobel-Prizewinning quantum physicist Erwin Schrödinger, and involves a hypothetical cat shut in a box with a bottle of poison, some radioactive material and a geiger counter. The apparatus is wired up so that, if the radioactive material decays, the geiger counter will be triggered and will set off a device to smash the bottle of poison and thereby kill the cat. If we set this experiment up, shut the lid of the box, and wait until there is a precise 50:50 chance that the radioactive decay has occurred, what, asked Schrödinger, is the state of the cat in the box *before* we open the lid to look?

Commonsense tells us that the cat is either alive or dead. But quantum physics tells us that events such as the radioactive decay of an atom become real only when they are observed. That is, quantum physics says that in this case the decay, or absence of decay, in the radioactive material is not decided until someone opens the box to take a look. Before we look in the box, the radioactive stuff exists in what is known as a superposition of states, a mixture of the decayed and not-decayed possibilities. Once we look, one of the options becomes real, and the other disappears. But before we look, everything in the box, including the cat, exists in a superposition of states. So the cat is described by quantum mechanics – a theory that has passed every test it has been subjected to, for more than half a century – as being both dead and alive at the same time.*

How can this be? One possible resolution of this puzzle goes by the name of the many-worlds hypothesis. It holds that whenever the Universe ('world', in this use of the term) is confronted by a choice of paths at the quantum level, it actually follows *both* possibilities, splitting into two universes (these are often described as 'parallel worlds', although in fact, mathematically speaking, they are actually at right angles to each other). On this picture, when the radioactive material in the box is faced with the choice of decaying or not decaying, it doesn't just dissolve into a ghostly dither of superposed states. Instead, the entire Universe splits into two. In one world, the material decays and you open the box to find a dead cat. In the other world, the material does not decay, and you open the box to find a live cat. Both cats, and both 'yous', are equally real, and neither has any knowledge of its counterpart in the other world.

The many-worlds interpretation of quantum mechanics is by no means taken seriously by all physicists. Intriguingly, though, among the minority which does take it seriously are some of the very best physicists of recent times, including John Wheeler (at one time, although he has since expressed doubts), Kip Thorne and Stephen Hawking (who thinks he can explain the origin of the Universe in a variation on the many-worlds theme). Such a possibility certainly resolves the granny paradox neatly – what happens is that the time traveller may go back in time and cause the death of poor old granny (or, rather, poor *young* granny), but this action then leads to the creation of a new branch to the world tree, a universe in which the time traveller does not exist and never has existed. When the time traveller moves forward in time again from the moment of granny's death, he or she will be moving up this new branch to the tree of time, arriving in a different world from the one they started from.

Science fiction has often explored this possibility. One of the most famous examples is in the novel *Bring the Jubilee* by Ward

* The detailed basis of this bizarre idea is explained in my book *In Search of Schrödinger's Cat*.

Moore. In that story, the main character initially lives in a world very much like our own, except that the South won the American Civil War. He travels back in time to study a crucial battle in the war, and inadvertently sets off a train of events which alters the course of that battle and ultimately leads to the victory of the United States over the Confederacy. When he travels forward in time again, he arrives in 'our' world. But his original world may still exist, on its own time track. The theme was also explored in the *Back to the Future* movies, most notably (if confusingly) in the second part of the trilogy.

So there are at least two ways in which time travel can occur without violating causality – if the causality is inviolably built in to the past, and if new universes can be created to accommodate any tinkering with past events. There is also another bizarre possibility – a time loop in which events are their own cause (or, if you prefer, something happens without a cause). Once again, science fiction provides a classic example.

Time loops, and other twists

In his story, 'All You Zombies', Robert Heinlein describes how a young orphan girl is seduced by (it turns out) a time traveller, and has a baby daughter who is left for adoption. As a result of complications uncovered by the birth, 'she' has a sex change operation, and becomes a man. Her seducer recruits her into the time service, explaining that she is, in fact, his younger self – and that the baby (which he has actually taken back in time to the orphanage where she grew up) is also their younger self.* The closed loop is delightful, and also violates no known laws of physics (although the biology involved is decidedly improbable). But what if we ignore such 'special effects', and also assume that nobody is daft enough to do anything which might create a paradox, like killing your own granny? How can we describe a simple piece of time travel in the language of modern physics?

The best way is to use a spacetime diagram. Imagine an inventor who works away in his laboratory, building a time machine. Once it is complete, he jumps in, flips a switch, and travels backwards in time and slightly sideways in space until he is sitting alongside his

* The story has been reprinted many times; see, for example, *The Best of Robert Heinlein 1947–59* (Sphere, London, 1973); my favourite full-length variation on this theme is *The Man Who Folded Himself* by David Gerrold (Faber, London, 1973).

younger self. Then, he switches off the machine, the two versions of the inventor exchange a few words, and eventually he proceeds on his way into the world outside the lab. The appropriate spacetime diagram of this pattern of events looks like Figure 7.1. In

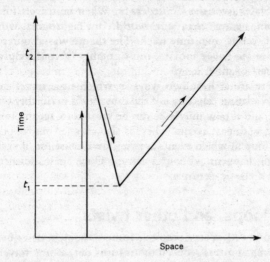

Figure 7.1 *Richard Feynman developed a variation on the spacetime diagram. In this example, the map shows how a time traveller completes building his time machine at time t_2 and travels back in time to talk things over with his earlier self at time t_1 before going back to the future.*

a slight variation on the standard spacetime diagram used by Minkowski, Richard Feynman developed the idea of using this to represent the flow of time. If you cut a narrow slit in a piece of paper or card, and place it over the diagram so that only the bottom axis is visible through the slit, you get a view of the location of the inventor in the laboratory at the moment he begins his work. Move the slit up the page (or just cover the diagram with your hand and move it up the page) and you see the world line of the inventor lengthening as time passes, but he stays in the same place. Suddenly, out of nowhere, an older version of the inventor appears, sitting in the time machine. From then on, for a time, we see *three* inventors. One, the youngest version, is building his time machine, having exchanged a few words with his older self. Another, the oldest, is going off into the world outside, having exchanged a few words with his younger self. And the third, sitting in the time machine, is of intermediate age. Not only that, but as time passes (moving *up* the page), he gets younger. We could tell this if, for example, he was smoking a cigar. From our God-like perspective outside of space and time, we would see the cigar start out as a stub between his lips, but growing longer as we move our focus of attention up the page, until it burns out into a complete cigar which the traveller carefully wraps up and tucks away in his pocket. What the time machine has done is to reverse the flow of time in its interior – an effect which is indicated by the fact that the world line for this third version of the inventor folds back on the inventor's initial world line.

Feynman's diagram was actually developed to describe the behaviour of particles in the subatomic world. A diagram like Figure 7.1 might usually be used to describe the appearance of a particle-antiparticle pair (perhaps an electron and a positron) at the 'V'. Although I have previously referred to such virtual pairs as annihilating with *each other*, and this is certainly by far their most common fate, in fact the equations can be balanced equally satisfactorily if one of them annihilates with a partner from the real world, repaying the energy debt to the vacuum so that the original virtual partner gets promoted into reality in its place. In that case, the positron from the virtual pair created at the 'V' in Figure 7.1 might soon meet up with an electron (the vertical line on the left), leaving its counterpart electron to scoot off into the world at large. Feynman caused consternation in the 1940s when he pointed out that this whole pattern could be regarded as the world line of a single electron that moves first forward in time, then backward in time, then forward in time again. A positron, in other words, is exactly the same as an electron travelling backwards in time.

You don't even need to invoke *virtual* particles to accomplish this trick. *Real* particle-antiparticle pairs can also be made out of pure energy, provided there is enough of it. When an electron and a positron annihilate, they release energy in the form of gamma rays; sufficiently energetic gamma radiation can also create a particle-antiparticle pair. So another version of a simple Feynman diagram might look like Figure 7.2. The implication is that in some sense all particle tracks and interactions may be fixed in the geometry of spacetime, with all movement and change being an illusion result-

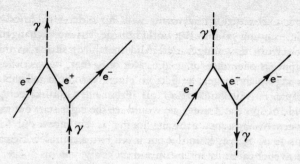

Figure 7.2 *When an electron-positron pair is made out of gamma radiation, the positron may annihilate with a different electron, leaving its original partner free. Feynman pointed out that this is exactly equivalent to a single electron bouncing off a gamma ray and travelling backwards in time before (?) bouncing off a second gamma ray and continuing its path into the future. Just as in Figure 5.7 (page 151), the laws of physics are entirely happy with particles travelling backwards in time. A positron, said Feynman, is an electron travelling backwards in time.*

ing from our changing psychological perception of the moment 'now' (Figure 7.3). Physicists have now become used to the idea, at

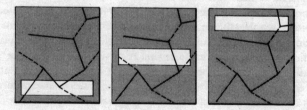

Figure 7.3 *Is time an illusion? If all particle world lines were somehow fixed in spacetime, and all that 'moved' was our perception, shifting 'up the page' as time 'passes', we would still see a complex dance of interacting particles, even though nothing was moving!*

least to the extent that Feynman diagrams are a valuable tool of the particle physics trade.★ But nobody 'really believes' that positrons are electrons travelling backwards in time – this is regarded as a metaphor rather than an expression of reality. Nevertheless, the laws of physics say that a positron is literally indistinguishable from an electron travelling backwards in time. And the fact that the world line on the same diagram can describe the adventures of a time-travelling inventor means that the laws of physics permit such jaunts (and, if you like, that an inventor travelling backwards in time is equivalent to an 'anti-inventor').

I have glossed over one point, which may be of practical importance when and if real attempts to build time machines are made. In the particle world, a particle-antiparticle pair may be manufactured out of gamma ray energy. But where has the mass-energy for the duplicate inventor come from? In order for an extra copy of the inventor to exist at the same time as the original, it seems reasonable to guess that the time machine will need an input of energy at least equivalent to the mass of the inventor. That would be a great deal of energy indeed; you won't get it by plugging your time machine into the electric mains (or even by making use of a convenient lightning strike), and this might restrict initial time-travel experiments to simple tests involving small amounts of matter, rather than whole human beings. But it is only a technological problem, and one rather less difficult than handling cosmic string. I never said that time travel would be easy – only that it is allowed by the laws of physics!

So let's stand back, for a moment, from the dramatic implications of *people* travelling in time. Concentrate, instead, on the notion of *particles* that travel backwards in time. This raises its own version of the basic time paradox, since if we had a means of sending particles back into the past then surely we could use them to send messages. Suppose that you and I have a time radio and an agreement along the following lines. I promise to telephone you a message at six in the afternoon, using the ordinary phone lines, provided that I have not received a time-radio message sent backwards in time by you to reach me at five o'clock. But you promise to send the message backwards in time provided that I *do* telephone at six. You *only* send your message to me if I phone; but I *only* phone you if you have *not* sent a message to me. Assuming we

---

★ I describe their use more fully in *In Search of Schrödinger's Cat*.

both keep our word,★ how do we resolve the dilemma? Now this, unlike the granny-murdering scenario or the time-travelling inventor, could become a real problem in the not too distant future. For, according to good old-fashioned relativity theory itself, and leaving aside Feynman's more recent conjuring tricks with spacetime diagrams, there is nothing wrong with the idea of particles that travel backwards in time. The only requirement is that they must *always* travel backwards in time – and, incidentally, that they travel faster than light. They even have a name – tachyons – although, perhaps fortunately, nobody has yet unequivocally found one.

Tachyonic time travellers

At first sight, the special theory of relativity seems to forbid faster-than-light (FTL) travel. If you start out moving slower than the speed of light and go faster and faster, time runs more and more slowly until, at the speed of light itself, it comes to a stop. You can't go any faster, because the speed of light itself is an impenetrable barrier – if you try to increase your speed any more, there is no time left in which to make the increase. But just on the other side of that barrier, according to the equations, lies a bizarre counter-clock world. There, if you are moving at just over the speed of light, time runs very slowly backwards. There is a certain logic to this – after all, if time runs slower as you approach the speed of light, and stands still at the speed of light, then it must run backwards ('slower than standing still') above the speed of light. The faster you go, in the tachyonic world, the more rapidly time runs backwards – and the more energy of motion such a particle has, the slower it goes (that is, adding energy always pushes a particle closer to the speed-of-light barrier, from either side of the barrier). So as a tachyon *loses* energy it goes faster and faster, rushing backwards in time as it does so. Amazingly, this bizarre possibility was first put forward just *before* Einstein published his special theory of relativity. At the beginning of the twentieth century, Arnold Sommerfeld (who had been a *Privatdozent* at Göttingen University, but was then a professor at the Technical Institute in Aachen and went on to gain fame in Munich as a pioneer of quantum theory) realized that

---

★ And it would be easy to get round that problem by setting up an automatic system that makes the phone call only if it receives the radio message, but sends the radio message only if it has not received the phone call.

Maxwell's theory of electromagnetism required FTL particles to speed up as they lost energy. He published this conclusion in 1904; since the special theory, published in 1905, is also largely based on Maxwell's theory, it is no real surprise that it contains the same kind of description of FTL particles. But nobody paid much attention to the idea until the 1960s, and even then it was regarded more as a game you could play with the equations than as a serious practical possibility. The hypothetical existence of such tachyons is another manifestation of the positive–negative symmetry inherent in many of the equations of physics, rather like the symmetry which allows for the existence of antiparticles. Nobody took the idea of antiparticles seriously either, when it was first propounded, dismissing this symmetry as a mathematical quirk of the equations. Now, antimatter is a routine part of physics, and is routinely manufactured in particle accelerators like those at CERN. But tachyons are not the antiparticle counterparts of known particles; they are (if they exist) a whole new possibility in their own right.

How could you ever spot a tachyon? The obvious place to look is in the showers of cosmic rays – particles from space that frequently smash into the top of the Earth's atmosphere. When an energetic cosmic ray particle collides with an ordinary atomic particle at the top of the atmosphere, it produces a shower of lesser particles which can be detected on the ground (indeed, this was the way in which positrons were first discovered). If some of the particles created in this way are tachyons, they will travel backwards in time, and arrive in the detectors on the ground not only before most of the particles in the shower but even before the original cosmic ray (the 'primary') hits the top of the atmosphere.

Cosmic ray investigators have scanned their records for traces of such precursor blips showing up in their instruments shortly before the arrival of conventional cosmic ray showers. They have found several blips that might fit the bill, but none of them offers unequivocal proof of the existence of tachyons, in spite of a flurry of excitement in the early 1970s. It was in 1973 that two researchers based in Australia, Roger Clay and Philip Crouch, found what seemed to be strong evidence of FTL precursor blips showing up in their cosmic ray detectors. Their results were sent to the journal *Nature*, where I was working at the time, and were published in 1974,* to, as I well recall, the consternation of many physicists and

* Volume 248, p. 28.

the delight of many journalists. Those results still stand, but are no longer regarded seriously as evidence of tachyons, because subsequent experiments have failed to find the precursors associated with other cosmic ray showers. It is generally accepted in the physics trade that something else must have set the Australian detectors off in 1973 at just the right (or wrong, depending on your point of view) time. But this wasn't quite the end of the search for tachyons.

Another way in which tachyons might make their presence known is if they (or at least some of them) are electrically charged. Einstein's speed-of-light limit refers, strictly speaking, to the speed of light *in a vacuum*. This is the famous constant c, for which no particle that is ever moving slower than c can be given enough energy to exceed the speed of light *in a vacuum*. But light itself moves more slowly than c when it passes through a transparent material such as a sheet of glass or a tank of water. 'Ordinary' particles can therefore travel faster than the speed of light in, say, water, without exceeding the ultimate speed limit c. When a charged particle, such as an electron, actually does this, it radiates light. Just as a fast-moving object that breaks through the sound barrier creates a sonic boom, so a fast-moving charged particle that breaks through the light barrier produces a kind of 'optic boom'. The effect was discovered by a Soviet physicist, Pavel Cherenkov, in 1934, and is known as 'Cherenkov radiation' in his honour. A charged tachyon, moving faster than light even in a vacuum, would also have to emit Cherenkov radiation, as long as it had any energy available to radiate. Calculations suggest that any such particle would lose all its energy literally in a flash, ending up with zero energy and travelling at infinite speed, so that in some sense it would be everywhere along its world line at the same time. If that world line intersected with another particle, however, the tachyon might thereby temporarily gain energy from the collision, and emit another flash of light. Alas, no appropriate flashes of light occurring in tanks of water have been detected, even though searches have been carried out at several laboratories.

The consensus is that real tachyons do not exist. They are, according to conventional wisdom, an artefact of the equations that can safely be ignored, regarded as having no real physical significance. Physicist Nick Herbert, of Stanford, sums the situation up neatly in his book *Faster Than Light*; 'most physicists', he says, 'place the probability of the existence of tachyons only slightly higher than the existence of unicorns.' And yet, they *are* allowed by the laws of physics, and one physicist, Gregory Benford, has used the idea to great effect in his novel *Timescape*, which also invokes the existence of parallel worlds. Even in Benford's fictional world(s), however, there is no physical transportation of ordinary objects (let alone people) backwards in time. If we want to achieve this trick, we are going to have to come up with some way of altering the structure of spacetime itself. Wormholes have obvious possibilities; but there is another possibility, which is, in some ways, simpler. It involves rotation, and it stems from the realization that if the whole Universe is rotating, then it is itself a time machine, in the sense that it contains closed timelike loops.

Gödel's universe

The man who came up with this idea had a habit of making disconcerting theoretical discoveries. He was the mathematician Kurt Gödel, who was born in 1906 in Brunn (which at that time was part of Austria; it is now in Czechoslovakia). He studied mathematics at the University of Vienna, and obtained his Ph.D. in 1930. Immediately after that, he produced a bombshell – a paper, published in 1931, that has sometimes been described as the most significant event in the study of pure mathematics in the twentieth century. Gödel showed, in a nutshell, that arithmetic is incomplete. If any system of rules is set up to describe simple arithmetic (and I really do mean simple; we are talking about two and two making four, here), there are bound to be arithmetical propositions, Gödel proved, that can neither be proved nor disproved using the rules of the system itself. This is now known as Gödel's Incompleteness Theorem. Now, it has to be said at once that this doesn't pose any problem in the everyday use of arithmetic. The rules of addition, subtraction and so on still work perfectly well, just as they did before 1931. But it is deeply worrying to logicians and philosophers, and it does mean that in principle it is possible that there might be something in mathematics that cannot be proved to be either true or false.

You can get a feel for what this means by looking at an old logical puzzle involving words, put forward in ancient times by the Greek philosopher Epeminides. He drew attention to the inherent logical inconsistency of self-referring statements such as the sentence:

This statement is false.

If the sentence is true, then it must be false; if it is false, then it must be true. You can ask the question 'Is the sentence true or false?', but that question has no answer. Such puzzles don't stop us using language effectively in everyday communication, and many ordinary people would dismiss any discussion of the meaning of such sentences as logical hair-splitting. But the important point, both with Epeminides' example and with Gödel's Incompleteness Theorem, is that self-referring loops can lead to logical contradictions – or, if you like, nonlogical contradictions. This has been used as the basis for arguing that, for example, human intelligence will never be able to understand the human mind, because in studying ourselves we are inevitably confronted with such logical loops. All of this forms a central theme of Douglas Hofstadter's excellent book *Gödel, Escher, Bach*; but it would be straying too far from my own theme to go any further into the fascinating implications here, except to point out that the existence of statements or mathematical propositions that cannot be proved to be either true or false does seem, in a sense, to echo the puzzles posed by time loops in which, for example, granny is both murdered and not murdered, and the quantum puzzle of the cat that is neither alive nor dead.

After the Nazis took over Austria at the end of the 1930s, Gödel himself moved to the United States, where he became a professor at Princeton, working alongside his close friend Albert Einstein. For a man who was able to prove logically that mathematics is incomplete, the equations of the general theory of relativity must have seemed a doddle, and, inspired by his friendship with Einstein, Gödel made several important contributions to relativity theory, finding new solutions to the equations. The most interesting of these variations on the relativistic theme emerged in 1949, when he came up with the idea that the natural tendency for gravity to pull the Universe together and make it collapse might be countered by centrifugal force if the entire Universe were rotating. Such a rotating universe would not have to have a unique centre to rotate around, any more than the expanding Universe possesses a unique centre from which it expands. In the Universe we see around us, any observer, wherever they are located, will see a uniform expansion apparently centred on the observer; in a similar fashion, in Gödel's universe any observer, wherever they are located, will see the universe apparently rotating about the observer. But that isn't all they would see.

When massive objects rotate, they drag spacetime around with themselves, in a manner reminiscent of the way coffee will swirl around if you twiddle your spoon in the cup. This happens very strongly in the ergosphere around a rotating black hole, and is the reason why strange processes which allow us (in principle) to extract energy from the black hole take place there. Indeed, the effect operates for *any* rotating mass, no matter how small – it's just that the dragging of spacetime is too tiny an effect to be noticed unless the rotating object is reasonably massive. It is, though, just possible that the effect may be big enough to be detected for the Earth. If this dragging of spacetime occurs as predicted by Einstein's general theory of relativity, it would show up as an influence on the way spinning gyroscopes behave in the vicinity of the Earth. The direction of the spin of the gyroscopes will change slightly, precessing because of the Earth's rotation. The predicted effect is tiny; but for two decades researchers at Stanford University have been working on a project to measure it. Their plan is to manufacture perfectly balanced gyroscopes, in the form of uniform spheres of metal, which will be flown into Earth orbit on board the Space Shuttle some time before the end of the 1990s and spun up under weightless conditions. There, a battery of instruments will watch the weightless gyros to see if they do indeed precess as a result of the dragging effect of the rotation of the Earth on nearby spacetime.

It is very difficult indeed to measure such an effect for such a small rotating mass as a planet. But if the whole Universe is rotating, similar effects should show up in a very dramatic way. The best way to get a picture of what is going on is in terms of the light cones which indicate the relationship between points in spacetime on a standard Minkowski diagram (not, this time, a Feynman diagram). Figure 7.4 shows the light cones associated with three points in spacetime, A, B and C. These points can know nothing about each other, and have no influence on each other, because for a signal to get from any of these points to either of the others it would have to pass outside the respective light cones, travelling faster than light. But as time passes, observers that start out at each of these points will follow their own more or less wiggly world lines into the future and up the page. At some time in the future, the observer who started at point A will receive light signals that come from point B, and this is the first time that such an observer can be influenced by events that occurred at point B. But this observer can never have any influence *on* events at point B,

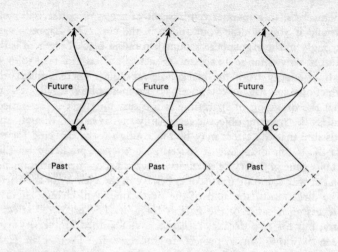

Figure 7.4 *A set of three light cones 'belonging' to spacetime events A, B and C. It is impossible to travel from any of these events to any of the others.*

because to send a signal there it would have to go backwards in time (in this discussion, I am assuming that tachyons do not exist); any interaction is strictly one-way. The same sort of pattern applies to the other observers, and, indeed, to all observers in flat spacetime.

But if the observers inhabit a universe that is rotating, they will find that it drags spacetime round in such a way that the light cones (everywhere in the universe) are tipped over. If it is rotating fast enough, the light cones tip over so much that an observer who starts from point A can get to point B without ever going outside the future light cone – that is, without ever exceeding the speed of light. An observer who starts out at point B can similarly visit point C, and we can imagine an overlapping set of light cones joining up to provide a circular route around the entire universe and back to point A (Figure 7.5). But this, remember, is a space*time* diagram. Point A represents both a location in space *and* a moment in time. In Gödel's universe, it is possible to set out from a point in spacetime and travel around the universe in a closed path that brings you back to the same place *and the same time* that you started from, even though the journey may have taken thousands of years according to the clocks carried along in your spaceship.

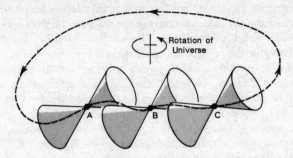

Figure 7.5 *If the Universe is rotating, the light cones may be tipped so that you can travel from A to B to C – and on around the Universe, back to event A. That is, back to the same place and the same time that you started from – and all without ever travelling faster than light.*

There, of course, lies the rub. In order to produce closed timelike loops in this way, a universe like our own would have to be rotating once every 70 billion years. This is a fairly leisurely rotation rate, for a Universe that is currently thought to be about 15 billion years old, and difficult to measure, although the available evidence is definitely against the Universe possessing this much rotation. Even if the Universe did rotate that fast, however, the shortest CTL would be about a hundred billion light years in circumference. That is, it would take a hundred billion years for even a beam of light to circle the Universe and get back to the same point in spacetime that it started from. Actually using such a universal time machine is not a practical possibility. But Gödel's solution to Einstein's equations indicates, yet again, that time travel is not forbidden by the general theory. It also shows that rotation, and the tipping of the light cones that it causes, can lead to the existence of closed timelike loops. In 1973, a researcher at the University of Maryland realized that you can do the same trick with much less mass than the entire Universe, provided that the matter involved is sufficiently compressed and is rotating very fast indeed.

Tipler's time machine

Frank Tipler, who came up with this dramatic idea, is now based at Tulane University, in New Orleans. He is a highly unconventional mathematical physicist, who as well as calculating how to build a time machine has a deep interest in the question of whether there is any other form of intelligent life in the Universe, apart from ourselves (for the record, his conclusion is that it would be so easy for any civilization slightly more advanced than ourselves to colonize the entire Universe that the fact that we do *not* see any signs of such a civilization in our astronomical back yard, the Solar System, can be taken as strong evidence in favour of the rather sobering conclusion that we represent the most advanced civilization there is). I first made contact with Tipler in 1980, when I wrote up his ideas about time travel for the magazine *New Scientist*, where I was working. We have kept in touch since, and he assures me that his calculations from the 1970s still stand up. His mathematical description of a workable time machine actually appeared in print in 1974, in the pages of the journal *Physical Review D* (Volume 9, pp. 2203–6), under the title 'Rotating cylinders and the possibility of global causality violation'.* To you and me, 'global causality violation' simply means 'time travel'. When I asked Tipler whether he seriously thought time travel possible, he replied that 'There is indeed a real theoretical possibility for causality violation in the context of classical general relativity.' And the methodical, thorough way in which he arrived at that conclusion provides a solid basis for any further speculation about time-travel possibilities.

Tipler spelled out his route to the mathematical blueprint for a time machine in three steps. First, he asked whether the equations allow in theory for the existence of journeys through spacetime in which the traveller returns to their starting point in both space and time, having travelled backward in time for part of the journey. We already know that the answer is yes – Gödel proved that in 1949, and there are other examples of solutions to Einstein's equations that allow CTL. Indeed, Brandon Carter showed in 1968 that Kerr's solution to Einstein's equations, describing spacetime in the vicinity of a rotating black hole, also contains closed timelike loops when the rotation is fast. Tipler knew about the earlier work, but, being cautious, he first established that CTL are permitted by the general theory to his own satisfaction. Then he asked whether it is possible for conditions under which journeys around closed time-

---

* Sf writer Larry Niven was sufficiently impressed to steal (with an acknowledgement!) not just the idea but the title from Tipler's paper for a short story, which can be found in the collection *Convergent Series* (Del Rey, New York, 1979).

like loops are possible to occur naturally in the Universe. The answer was, again, 'Yes'. Finally, he asked himself whether it is possible, in principle at least, to create such conditions artificially – that is, to build a working time machine. Once again, the answer was 'Yes'.

The key feature in Tipler's calculations, presented in the 1974 paper and in later work, is rotation. But he also found that a time machine of this kind (natural or artificial) cannot be created from ordinary matter under ordinary conditions; you have to have a rotating naked singularity in order to have closed timelike loops. As far as nature is concerned, as we have seen, this possibility is by no means ruled out, since naked singularities may form when black holes explode, or when non-spherical aggregates of matter collapse under the pull of gravity – and in either case it would be astonishing if the end products were not rotating. But by far the most interesting aspect of Tipler's work is his description of the basics of an artificial time machine.

The way light-cone tipping leads to time travel is shown in Figure 7.6. In this version of a Minkowski diagram, *two* space dimensions, X and Y, are indicated, with the flow of time, as usual, going up the page. Only the future halves of the light cones are shown, to keep the picture simple. The time axis also represents the world line of a massive, rapidly rotating naked singularity, wrapped in a strong gravitational field; the interesting effects on light cones are shown by looking at lines that trace circular paths around the singularity at different distances. Far away from the singularity, where the gravitational field is weak, the light cones open out into the future in the usual way for flat spacetime. But the closer you get to the spinning singularity, the more the cones tip over, in the direction that the central object is rotating. For an observer in such a situation, everything would appear normal, and, for example, the rules of special relativity restricting travellers to velocities less than that of light still hold. But to an observer far away in flat spacetime, watching events in the region of distorted spacetime, the roles of space and time in that strong field region can be seen to begin to interchange. Time itself begins to twist around the central object.

The critical stage for the light-cone tipping, as far as time travel is concerned, is when the cone is tipped by more than 45°. Since the half-angle of the cone is 45°, this is the point where the future light cone tips so far that one edge of it lies below the XY plane

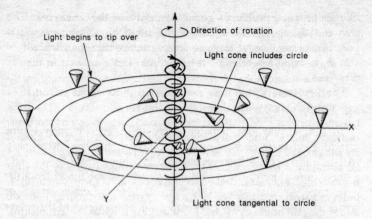

Figure 7.6 *A massive, rotating cylinder will also drag spacetime around with it and cause the light cones to tip over in the region of strong gravitational field. This is the basis of Frank Tipler's design for a time machine. By travelling in a tight orbit around the rotating cylinder, you would travel backwards in time, as represented by the central helix in this diagram.*

representing all of space. Part of the future light cone in the region of strong field now lies in the *past*, as viewed from the region of weak field. Remember that a space traveller can, in principle, go anywhere within the future light cone. In this extreme tipped-over light-cone situation, the traveller can choose to move along a path which, to the outside observer, consists solely of a circle around in space, without any motion through time (up the page) at all! In some sense, the traveller would be everywhere around that orbit at the same time. And if the traveller chose to steer a spaceship on a course that dipped just below the XY plane, it could travel in a gentle spiral round and round the time axis, gradually moving *down* the page and backwards in time, indicated by the spiralling 'orbit' shown in the middle of Figure 7.6. The spaceship would keep returning to the same place, but at earlier and earlier times. Then, by a judicious adjustment of the orbit, the traveller could follow a similar helical path forwards in time and back to the future. As Tipler puts it:

A traveller could begin his journey in weak field regions –

perhaps near the Earth – go to the tipped-over light cone region and there move in the direction of negative time, and then return to the weak field region, without ever leaving the region defined by his future light cone. If he travelled sufficiently far in the minus-t direction while in the strong field, he could return to Earth before he left – he can go as far as he wishes into the Earth's past. This is a case of true time travel.

In fact, it may not actually be possible, even if such a time machine exists, to go back as far as you wish into the Earth's past. All of the effects I have described, involving tipped-over light cones, apply only to the region of spacetime in the future from the spacetime point at which the time machine (whether natural or artificial) is created. Such a time machine opens up all of the future of spacetime to exploration; but it is impossible, using such a machine, to go back further in time than the moment the machine itself was created. This means that if we were to build a time machine tomorrow, we could not use it to go back to study the way the ancient Egyptians build the pyramids. That is only possible if a time machine already existed then, and we are lucky enough to find it and learn how to use it. Some time-travel enthusiasts seize on this as an explanation for why we have not yet been visited by time travellers. They suggest that the reason is not that time travel is *impossible*, as other people argue, but simply that no time machine has been invented yet! Even enthusiasts are slightly disappointed, though, that there is no prospect of building a time machine tomorrow and using it to hop back to interesting events in Earth history. However, there is a compensating bonus involved in the creation of a Tipler time machine. It only has to exist for an instant in order to open up the entire future for exploration, because the closed timelike loops tied to the time machine extend into the entire future from the moment that the machine is created. But the key question remains: How would you set about building such a device at all?

The best prospect, in principle, is to find a very compact, rotating object that has been produced naturally in the Universe, and to speed up its rotation to the point where closed timelike loops form around it. What we need is a massive, compact, fast-spinning cylinder. The best place to start is with a neutron star. Neutron stars are the most compact, dense objects known, and some of them also spin very fast. At least one pulsar is known that spins on its axis once every 1½ milliseconds (it is known, with slight exaggeration, as the 'millisecond pulsar'). This is surprisingly close to the rotating speed at which a natural time machine might form, according to Tipler's calculations. He says that if a rotating massive cylinder is spinning fast enough, then a naked singularity will form at its centre, with closed timelike loops tied to that singularity. Such a cylinder would have to be at least 100 km long, and no more than 10 to 20 km across, containing at least as much mass as our Sun and with the density of a neutron star, with the whole thing rotating *twice* every millisecond – only three times as fast as the millisecond pulsar. Indeed, if you took ten neutron stars, joined them pole to pole, and gave them enough spin, you would have a Tipler time machine.

Of course, there are enormous problems involved in such an engineering feat, not least being where you find ten neutron stars to start with. The rim of the cylinder would be moving in a circle at half the speed of light, and the energy associated with the strong angular momentum of this rotation would be about the same as the rest-mass energy (the 'mc^2') of the cylinder – 'energy so great', says Tipler, 'that the accompanying centrifugal force may tear the rotating body apart.' And while the cylinder is trying to tear itself apart in one direction, it is trying to collapse in the other, along its length. The gravitational pull of ten neutron stars joined end to end would quickly make them collapse into a black hole, unless some form of energy field stronger than anything we yet have any direct experience of could hold the cylinder rigid. It sounds just about impossible – but remember that the singularity only has to form for the most fleeting instant in order to provide the closed timelike loops that would make time travel possible for ever afterwards. Like so many relativists before him, Tipler seems to be telling us that time travel is indeed possible in principle, but that the practical difficulties associated with building a time machine are enormous, and may be insurmountable. Nevertheless, I find the existence of millisecond pulsars tantalizing and intriguing – a classic example of 'so near, and yet so far.' Such objects are *so close* to being natural time machines that it is hard to resist the speculation that nature may already have done the job that human engineers would find so difficult. It seems to me more likely that our descendants will *discover* a pre-existing time machine (with the bonus that they then really·could use it to go back into history) than that they will build one.

But this is not the end of the story of time-machine engineering. The collapse of Tipler's would-be time machine into a black hole, and the hint that we may need fields that can hold things rigid with a force greater than anything known on Earth, seem to be pointing us back towards wormholes and cosmic string. If cosmic string exists, it would be the ideal stuff to thread Tipler's neutron stars on and stop them collapsing, just as it is the ideal stuff to hold open a star gate manufactured out of a wormhole. And as Thorne, Novikov and their colleagues have shown, once you have a wormhole that operates as a star gate, providing a shortcut through hyperspace, it is a simple matter, in principle, to convert it into a time machine.

Wormholes and time travel

Carl Sagan's simple request for a reasonably plausible piece of hokum with which to entertain the readers of his novel has caused ripples that have now spread far through the physics community, and right around the world. Novikov had been interested in the implications of the possible existence of CTLs for many years, and when the CalTech group began to appreciate that the kind of star gate they had invented to fit Sagan's fictional needs could also be used as a time machine, it was natural for Thorne to get in touch with Novikov, and for Novikov's team, based in Moscow, to get involved in trying to find out whether the laws of physics can handle the existence of CTLs in what Thorne calls 'a reasonable way'. The group directly involved in this research consisted (the last time I counted all the names on one of their scientific papers) of seven researchers based on two continents. Thorne has taken to referring to them as 'the consortium'; and there are others, including the Newcastle group, Ian Redmount (at Washington University, St Louis) and Matt Visser, who are equally interested in the implications. Most of the rest of the discussion in this chapter is based on the work of the Soviet-American consortium – starting with their technique for turning a star gate into a time machine.

Once you have a working wormhole star gate, you don't even need general relativity to tell you how to turn it into a time machine. The special theory is quite adequate for the task. Remember that if there are two identical twins, one of whom stays at home while the other goes away on a journey at a sizeable fraction of the speed of light and then returns home, the twin who goes on the journey will age less than the twin who stays at home. Moving clocks run slow. Given the engineering resources of a superior civilization, we can imagine catching hold of one mouth of the wormhole in some way, and taking it off on just such a journey. Of course, it isn't easy to get hold of something as nebulous as the mouth of a wormhole, but there are two obvious ways in which this might be done. First, it is one of the key characteristics of such a wormhole mouth that it has a large mass and a correspondingly strong gravitational field – it must have, in order to distort spacetime sufficiently to make an opening into a wormhole big enough for people and spaceships to travel through. All you need to attract a gravitating body is another gravitating body; it is possible to imagine dangling a large mass (perhaps a planet) in front of the wormhole mouth, and moving the large mass away so that the wormhole mouth follows along behind, like the proverbial donkey trotting along after a carrot on a stick that is always held just out of its reach. Alternatively, we can imagine adding a judicious amount of electric charge to the wormhole mouth (not enough, of course, to upset the geometry of the throat), and towing it with the aid of an electric field. No doubt a superior civilization will have other tricks up its sleeve, but these will do for now.

Once you have a means to tow one end of the wormhole around, you can take it on a long journey at close to the speed of light, then bring it back to rest alongside the other end of the wormhole. This could be a journey out to another star and back, or it could simply involve whirling the moving mouth around in a circle until you had built up a sufficiently impressive time difference between clocks in the moving frame of reference and clocks attached to the mouth that stays at home. What matters is that, even when the moving mouth has been brought back to rest, that time difference remains. It is a real, physical property of the region of space associated with the moving mouth; this has aged less than the mouth that hasn't moved, and is therefore in the past of the mouth that stayed at home.

Because of the way in which spacetime is connected by the wormhole geometry (the topology of spacetime associated with the wormhole) this means that the wormhole will act as a time machine. A traveller who jumps into the mouth that has moved will emerge from the stationary mouth *at the time corresponding to the time on the clocks of the moving mouth*. Suppose that the moving mouth has travelled far enough, and fast enough, to establish a time difference of one hour between the two mouths. A traveller who starts out from the stationary mouth when clocks there read twelve o'clock and takes, say, ten minutes to cross over to the moving mouth* will arrive there when the traveller's watch and clocks at the stationary mouth both read 12.10. But if the traveller now jumps into the moving mouth, when he or she emerges from the stationary mouth (almost instantaneously, according to the traveller's own watch), the time there will be 11.10. The traveller can now cross quickly over to the moving mouth, arriving there at 11.20, and jump in again, emerging from the stationary mouth at 10.20. And the whole sequence can be repeated, jumping back in time again and again, back to the moment when the time difference between the two ends of the wormhole was established. Like Tipler's time machine, the wormhole variety allows travel into the past only as far back as the time when the machine was created; but, also like Tipler's time machine, it allows indefinite travel into the future, in this case simply by entering the stationary mouth and emerging from the moving mouth a split second later by your watch but an hour later as far as the outside Universe is concerned.

The big practical difficulty is that you have to move the mouth far and fast in order to build up a useful time difference. Even travelling at 99.9 per cent of the speed of light for ten years before being brought to a halt will only slow the ageing of the moving mouth by nine years and ten months, creating a time difference of nine years and ten months between the two ends of the wormhole. But the practicalities are not the main concern of physicists who study the theory of time travel today – Kip Thorne has said (perhaps erring a little on the pessimistic side) that even if the laws of physics do allow for the construction of time machines the chance of building one within the next thousand years is 'nil'. What he and the rest of the consortium (and others) are concerned about is how to find, within the framework of the laws of physics which say time travel is possible, a logical set of equations that remove the physical basis for the famous paradoxes of time travel. If time travel really is possible, how do you avoid violating causality? Or, to put it another way, how can you doctor the paradoxes?

---

* The mouth has now *stopped* moving, of course, but this is still a convenient name to apply to it, and 'mouth that was moving' is too much of a mouthful.

Paradoctoring the paradoxes

There are two key features of the consortium's approach. First, they have nothing to do with problems involving human beings, who might change their minds about what they are planning to do, or tell deliberate lies about whether or not they intend to murder granny. This is fair enough, since the problems they are interested in are those that concern the basic physics of time travel, which is complicated enough without introducing the further complication of human psychology. There will be ample time to start worrying about the role of wilful human observers if and when we are happy that we understand the basic physics. So, in the tradition of using the simplest possible physical systems to highlight the underlying truths inherent in the equations, the consortium studies the way in which billiard balls might become involved in interactions with themselves when they travel through time tunnels.

The second basic feature of the consortium's attack on the time paradoxes is to assume that the Universe will only allow those solutions to the equations that are self-consistent. Again, this is quite reasonable, on two grounds. If inconsistent solutions are allowed, then all bets are off, and there is no point in trying to understand the basic physics; furthermore, it is quite common, even in simple, everyday physical systems, to find solutions to the relevant equations which are allowed mathematically but are physically impossible, and can be ignored. This often happens in the case of equations involving square roots. For example, Pythagoras' famous theorem about triangles, expressed as an equation, actually tells us that the lengths of the sides of a triangle could be negative; but we know that this 'solution' is physically impossible (there are no triangles in which, say, two sides are respectively 3 metres and 4 metres long, while the third side is −5 metres long), and ignore it. Similarly, the consortium assumes that only solutions to the equations of time travel that are 'globally self-consistent' are acceptable.

We can see what all this means, and how it provides new insights into the workings of the Universe, by looking at the billiard-ball equivalent of the granny paradox. We do this by imagining a time tunnel set up with its two mouths close together. If a billiard ball is fired into the appropriate mouth of the time tunnel in just the right way, it will emerge from the other mouth in the past, and just have time to travel across the intervening space to collide with itself before it enters the tunnel, knocking the earlier version of itself out of the way. So it never travels through time, the collision never takes place, and therefore the earlier version of the billiard ball *does* enter the time tunnel . . . and so on. This is the self-*inconsistent* solution to the problem, and the consortium says that it must be rejected – the Universe cannot possibly operate like that.

The reason why they are confident that it is acceptable to dismiss the self-inconsistent solution is that they have found that there is always another solution of the equations that gives a self-consistent picture starting from the same initial circumstances. Extending the analogy with Pythagoras' theorem, if there were only one solution to the equations, and it said that the length of one side of the triangle had to be negative, we would have to accept this at face value, even if we did not understand what it meant; but because there are two solutions, and because we understand all about triangles with sides that have lengths measured in positive dimensions, we can accept the physically meaningful solution and ignore the other one. In the same way, the consortium accepts only self-consistent solutions to their time-travel problems, and ignores the others.

An example of a self-consistent solution to this kind of billiard-ball problem is when the ball approaches the time tunnel and is struck a glancing blow by an identical billiard ball that has just emerged from one mouth of the time tunnel, knocking the first ball into the other mouth of the tunnel. As the first ball emerges from the other mouth of the tunnel, it collides with the younger version of itself, knocking itself into the tunnel (Figure 7.7). Thorne, Novikov and their colleagues have not only found that there are no billiard ball problems of this kind which do not have at least one self-consistent solution, but that every problem of this kind that they can think of has an *infinite number* of self-consistent solutions. Figure 7.8 shows how this can arise. In this case, we have a billiard ball that passes neatly in a straight line between the two mouths of the time tunnel. Or does it? Suppose that when the ball is midway between the two mouths it is struck a violent blow by a fast-moving ball that emerges from the stationary mouth. The 'original' ball is knocked sideways, travels through the tunnel and becomes the 'second' ball – but in the collision it is deflected back on to exactly the same path, or trajectory, that it was following before the collision. As far as any distant observer is concerned, it still looks as

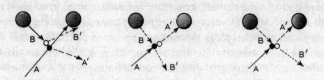

Figure 7.7 1. *The billiard-ball version of the granny paradox. If a ball (A) goes into one mouth of a wormhole time tunnel, emerges from the other mouth (B) in the past, and knocks its original self off course, how did it ever get into the wormhole in the first place?*
2. But if the 'second' ball bounces off the 'first' ball and into the hole in its place, there is no problem.
3. Nor is there a problem if it was the 'original' collision that knocked the ball into the hole in the first place!

if the single ball has passed smoothly in a straight line between the two mouths; and you can imagine similar patterns involving two, three or more circuits by the ball around the time tunnel. There seems to be more than one acceptable way to describe the ball's behaviour.

All of this is reminiscent of the way the Universe operates at the quantum level. There is a choice of realities, just as there is in the famous example of Schrödinger's cat. The billiard ball seems to be

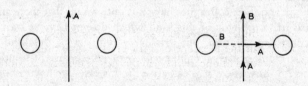

Figure 7.8 *In fact, there is an infinite number of 'self-consistent solutions' to the 'paradox', in which the ball can go round the loop many times in many different ways. From a distance, it looks, in this particular example, as if the ball has simply gone straight through the gap between the two mouths of the time tunnel. By averaging out the many different time-travel possibilities, the Universe arrives at a seemingly simple version of reality.*

perfectly normal before it gets near the time tunnel, then interacts with the tunnel system in many different ways, forming a super-position of states, before it emerges on the other side behaving, once again, in a perfectly normal fashion. What Thorne calls the 'plethora' of self-consistent solutions to the same billiard-ball/wormhole problem would be deeply troubling if it were not for the fact that quantum theorists have already worked out how to handle such multiple realities.

The technique they use was first developed by Richard Feynman in the 1940s, and is known as the 'sum-over-histories' approach. In classical physics – the physics of Newton – a particle (or a billiard ball) is regarded as travelling along a definite path, a unique world line, or 'history'. In quantum physics, there is no such thing as a definite trajectory, because of quantum uncertainty. Quantum mechanics deals only in probabilities, and tells us, with great precision, how likely it is that a particle will travel from one place to another. *How* the particle gets from one place to another is a different matter; the probability that tells us where the particle is likely to turn up next can actually be calculated by adding up probability contributions from all possible paths between the starting position and the end position. It is as if the particle is aware of all the possible routes it might take, and decides where it is going on that basis. Since each trajectory is known as a 'history', this technique of calculating how particles will behave by adding up the contributions from each trajectory is known as the 'sum-over-histories'.

Of course, all of this applies down at quantum level, on the scale of atoms and below. Quantum uncertainty is very small, and has a negligible influence on our everyday world, so that real billiard balls, for example, behave just as if they are following classical trajectories. But the presence of a traversable time tunnel in effect creates a new kind of uncertainty, in the region between the mouths of the tunnel, operating on a much larger scale. The consortium has found that the sum-over-histories approach works perfectly in this new situation, describing solutions to problems involving billiard balls that travel through time tunnels. If you start out with an initial state of the ball as it approaches the time tunnel from far away, then the sum-over-histories approach gives you a unique set of probabil-ities which tell you when and where the ball is likely to emerge on the other side, clear of the region containing closed timelike loops. It doesn't tell you how the billiard ball gets from one place to another, any more than quantum mechanics tells you how an electron moves within an atom. But it does tell you, precisely, the probability of finding the billiard ball in a particular place, moving in a particular direction, after its time tunnel encounter. What's more, the probability that the ball starts out moving along one classical trajectory and ends up moving along a different one turns out to be zero. As shown in Figure 7.8, from a distance an observer will not see the ball to have been deflected at all by its encounters with itself, and unless you look closely you will not notice anything peculiar going on. 'In this sense,' says Thorne, 'the ball "chooses" to follow, in each experiment, just one classical solution; and the probability for following each of the solutions is predicted uniquely.'* And there is a bonus. In the sum-over-histories approach, strictly speaking we are not ignoring the self-inconsistent solutions, after all. They are still there, in the addition of probabilities, but they make such a tiny contribution to the overall sum that they have no real influence over the outcome of the experiment.

There is one more very strange feature of all this. Because the billiard ball is, in some way, 'aware' of all the possible trajectories – all the possible future histories – open to it, its behaviour anywhere along its world line depends to some extent on the paths open to it in the future. Because there are many different paths that such a ball can follow through a time tunnel, but far fewer that it can follow if there is no time tunnel to pass through, this means that it will behave differently, in principle, if it has a time tunnel to go through than if it has not. Although it would be very difficult indeed to measure such an influence, according to Thorne this means that it ought to be possible, in principle, to carry out a set of measurements on the behaviour of billiard balls *before* any attempt to construct such a time machine has been made, and work out from the results whether or not a successful attempt to construct a time tunnel involving CTLs will be made in the future. This, he says, is 'a quite general feature of quantum mechanics with time machines'.

Summing up the work of the consortium to date, Thorne comments† that the behaviour of the laws of physics in the presence of time machines seems to be sensible enough 'to permit physicists to continue their intellectual enterprise without severe dislocation', even though time machines seem to endow the Universe with

* CalTech preprint number GRP–251.
† op. cit.

'features that most physicists will find distasteful.' It *is* possible to construct time machines, according to the laws of physics, and it *is* possible to have time travel without violating causality. As Novikov put it in a talk at Sussex University in 1989, 'if there is a non-self-consistent solution to the problem and there is also a self-consistent solution, then nature will choose self-consistency.'

Even this, though, is not quite the end of the story of black holes and the Universe. Among the minority of physicists who do not seem to find these ideas distasteful there is a growing band of researchers investigating the way in which much smaller worm-holes than anything I have discussed so far may exist as a spacetime 'foam' at the quantum level. One reason why such 'microscopic' wormholes are intriguing is that, if they exist, it might be possible to make a time machine by capturing a microscopic wormhole and somehow expanding it up to macroscopic size. But this trick pales into insignificance alongside the possibility that microscopic worm-holes may actually explain the very existence of the Universe itself. Once again, the explanation involves Feynman's sum-over-histories technique.

CHAPTER EIGHT

Cosmic Connections

Baby universes and spacetime bubbles. Inflating the universal bubble. The big fix – goodbye to Einstein's embarrassing constant. Black holes and the ultimate fate of the Universe – the end of time, or time without end?

Quantum uncertainty doesn't just affect particles and energy in the Universe. It affects the structure of spacetime itself. The way to picture this is to go back to the old image of the expanding Universe as the skin of an expanding balloon, like the one shown in Figure 8.1. This, of course, is the view of the Universe from the

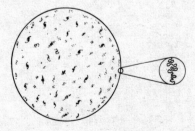

Figure 8.1 *Instead of being smooth like the skin of a balloon, the spacetime of the expanding Universe is a froth of quantum activity.*

perspective of some God-like observer standing outside of space and time. It looks, on this scale, very smooth and uniform, with a clearcut boundary. But now imagine a close-up view of a tiny portion of the skin of the balloon. If this could be magnified to show what is going on on a scale far smaller than the size of an atomic nucleus, down around 10^{-33} cm (the Planck scale) this hypothetical observer would see that spacetime itself was a constant seethe of activity, as turbulent as the surface of a storm-tossed ocean, wriggling about in an unpredictable manner, bending first one way and then the other. This is the effect of quantum uncertainty, exactly equivalent to the way in which virtual particles seethe through the vacuum.

One possibility – quite a likely one, according to researchers such as Stephen Hawking – is that during this seething activity a tiny wormhole will form in the fabric of spacetime on this scale. This might be a wormhole with both mouths 'in' our Universe, like the wormholes discussed by the Thorne-Novikov consortium but on a tiny scale. Scientists interested in time travel point out that in the far distant future a superior civilization might be able to capture one of these tiny wormholes and stretch it in some way, to create the kind of time tunnel that I described in the previous chapter. But there are other forms of quantum wormhole that are allowed by the equations of uncertainty – it might be a wormhole that pinches off a tiny piece of spacetime from our Universe, in such a way that the pinched-off portion of spacetime begins to expand and to form another universe in its own right, connected to our Universe by the microscopic wormhole (Figure 8.2). It is as if a blister developed on

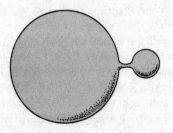

Figure 8.2 *Sometimes, a baby balloon may pinch off from the quantum 'skin' of the parent Universe.*

the surface of the balloon, pinched itself off from the main balloon and expanded independently.

Cosmologists refer to this possibility as a 'baby universe'. The only connection with the 'mother' Universe is through a wormhole whose entrance is a black hole about 10^{-33} cm across, which we would never notice. But the possibility of baby universes completely changes our understanding of the nature of our own Universe.

Blowing bubbles

The first important insight concerns the way in which the baby universe starts to expand. This is directly related to the way in which our own Universe started to expand away from the initial singularity. The process is known as inflation, and our understanding of it is based on a theory developed in the 1980s by Alan Guth of MIT. Inflation explains how a tiny seed of a universe, perhaps no bigger than a quantum fluctuation of the vacuum, can be blown up into the fireball of the big bang, literally in a split second of time. I have described the inflationary scenario in detail in my book *In Search of the Big Bang*, and I won't go into those details again here. The main point, though, is that before the idea of inflation came along cosmologists were happy that they could explain (essentially using the general theory of relativity alone) how the Universe got to be the way it is *starting out from the big bang fireball*; but they had no idea how that fireball of energy came into being in the first place. Inflation takes on board ideas from quantum theory to provide a natural mechanism which whooshes a microscopic seed of a universe, down on the Planck scale, up to the hot fireball stage, where relativity theory takes over.

The idea was developed in an attempt to explain the existence and nature of our own Universe. But, of course, if the trick works once then it can work repeatedly. Any microscopic, quantum fluctuation of the vacuum has the potential to be inflated into a new universe in its own right – although it is not inevitable that *all* quantum fluctuations will be inflated in this way (most of them probably just disappear, like virtual particles), *some* will form baby universes, and many of those baby universes will grow up into fully fledged universes comparable to our own. According to researchers such as

Guth and Hawking, this may be going on all the time, throughout our own Universe.

This raises the second intriguing and important point about the notion of baby universes. *Where*, exactly, is all this going on?

Remember that the surface of the balloon in Figure 8.1 does not represent the 'edge' of the Universe. The infinitesimally thin skin of the balloon represents *all* of space. So the quantum fluctuations in the structure of spacetime are actually going on everywhere throughout the three spatial dimensions of our Universe. When spacetime wriggles about, pinches off a blister, and forms a wormhole connection between the blister and the mother Universe, that blister exists in its own set of dimensions, all of which are at right angles to all of the dimensions of our Universe. Which means that the entire baby universe has no physical influence on our Universe, except through the wormhole, and cannot be seen or felt. A baby universe could be pinching itself off and inflating to become a real, grown-up universe in the room you are sitting in, right now, and you would never notice, because the only evidence that the new universe existed would be the mouth of a quantum wormhole, a black hole far smaller than a proton, making a tiny pucker in the fabric of spacetime somewhere in your room.

What all this means, of course, is that our Universe may have been born in the same way, as a blister in the spacetime of another universe. The overall structure of spacetime may be a kind of froth of expanding and collapsing bubbles, connected by wormholes, with no overall beginning and no end, extending to infinity in all directions, but in which individual bubbles (individual universes) may be born, expand for a time, and then collapse back into the froth (Figure 8.3). But none of this would be more than an exotic byway of speculative science, something to titillate the Sf fans but hardly to be taken seriously, if it were not for a dramatic discovery made at the end of the 1980s. According to some versions of the baby universes idea, information may leak from one universe into another through the microscopic wormholes that connect them. If it does, this could resolve one of the longest-standing puzzles in cosmology, the vanishing of the cosmological constant. And if that idea is correct, then, just maybe, the information leakage between universes might also account for why other constants of nature, such as the one that determines the strength of the gravitational force, and the one which fixes the amount of electric charge on an electron, have the values that they do.

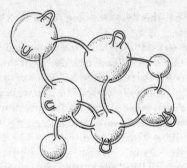

Figure 8.3 In fact, our Universe may be one of many spacetime bubbles connected by wormholes.

Einstein's vanishing constant

It was in 1987 that Hawking proposed that the existence of microscopic wormholes might alter the workings of quantum mechanics. At first, he thought that this would change the constants of nature in unpredictable ways, making it impossible ever to get a proper understanding of how physics works at this fundamental level. But just a year later, in 1988, Sidney Coleman, of Harvard University, suggested that just the opposite might be the case. In two key scientific papers, he argued that rather than making quantum mechanics unpredictable at the most fundamental level, it might be the wormholes themselves that actually fix the constants of nature.

The vanishing of the cosmological constant is the best example of this. Einstein brought the constant into his equations in order to hold the Universe steady, preventing it from either expanding or contracting, even though the raw version of those equations said that it must be doing one or the other. One way to think of the constant is as a kind of antigravity (Einstein's main concern was to stop the Universe collapsing under the influence of gravity, so he needed something to oppose the force of gravity), or as an energy possessed by the vacuum itself. The discovery, at the end of the 1920s, that the Universe really is expanding removed this motivation for the constant, because the observed expansion exactly matches the kind predicted by the equations of the general theory without such a constant. Indeed, if we had the effect of Einstein's cosmological constant added in as well, the Universe would expand much more rapidly than we actually see (if the constant were negative, it would make the Universe expand less rapidly than we actually see).

But in the 1980s interest in the cosmological constant was revived by the discovery that the nature of the expanding Universe can best be explained if it actually did undergo a phase of much more vigorous expansion, called inflation, during the first split second of the outburst from a singularity. This rapid expansion, which created the big bang fireball, is thought to have been driven by the negative pressure of a strong vacuum energy that existed at that time – in effect, by a positive cosmological constant. It is exactly this negative-pressure state of the vacuum that might be frozen into pieces of cosmic string left over from the inflationary phase, offering a chance for a superior civilization to obtain the scaffolding required to hold open the mouth of a traversable wormhole, in accordance with Matt Visser's blueprint. Once Alan Guth had come up with the idea of inflation, physicists found that there was no difficulty in providing the vacuum energy required from quantum processes. Indeed, one of the attractions of the idea is that this energy appears naturally from the quantum description of the early Universe. But the problem they were left with was what had happened to the cosmological constant. How had it managed to vanish so utterly at the end of the inflationary phase?

The size of this problem is best seen by thinking in terms of the Planck scale. This is in effect the quantum of length, the shortest distance that has any meaning at all. Quantum uncertainty fuzzes out the structure of space on any shorter scale. This minimum length is about 4×10^{-33} cm; that is, a decimal point followed by 32 zeroes and a 4 centimetres. It happens that the size of the cosmological constant can also be expressed in terms of length, because (like gravity) it is a measure of how the force between two objects varies as the distance between them varies. The way the Universe is seen to be expanding today shows that the cosmological constant must now be small even compared with the Planck length itself. It is very hard to see how any force could be *that* small without actually vanishing entirely. And wormholes explain how that vanishing trick could have happened.

Like gravity, the cosmological constant is a creature of geometry. Remember 'space tells matter how to move; matter tells space how to bend'. If you have an understanding of the overall geometry of the Universe, in terms of bent spacetime, then you have an understanding of the expansion, including the effects both of gravity and of vacuum energy. But according to the wormhole idea, the geometry that you have to understand is not just that of our expanding Universe, but of all the universes that are connected together by wormholes – sometimes called the 'meta-universe'. It is, of course, impossible to work out just what the geometry of the meta-universe is. But by applying the rules of quantum physics to the calculation of spacetime geometry, researchers such as Hawking and Coleman believe that they can tell us what *kind* of geometries are permitted.

This is where the many-worlds idea, and Feynman's sum-over-histories approach, come in. When we are thinking about individual particles moving from one place to another, Feynman's approach is to add up the probabilities of all the different possible routes the particle might take, in order to work out how likely it is that the particle really will go from one place to the other. When we are dealing with gravity, however, the important quantity (which in a sense corresponds to the position of a particle at any instant of time) is the entire geometry of three-dimensional space at some instant of time. The history of the Universe can be described as the evolution of geometry – the changing shape of the Universe – from one instant to another, just as the trajectory of a particle can be described as its motion from one point to another – its changing position in the Universe. So the idea behind quantum gravity is that it ought to be possible to describe the actual evolution of the Universe by adding up, in the correct quantum mechanical sense, all of the possible ways in which space can evolve from one three-dimensional geometry to another – including all of the possible wormhole geometries linking the meta-universe.

Now, this is still very difficult. But by making some simplifying assumptions (one of which is to tackle the problem in terms of a four-dimensional geometry of space, instead of three dimensions of space married up with one of time) the theorists believe that they can pin down some of the general properties that any expanding bubble within the meta-universe would have. In particular, information about the laws of nature leaks into each universe (including our own) from its neighbours, through the wormholes. And if any bubble starts out with a non-zero cosmological constant, it turns out that interactions taking place through the wormholes produce an effect that is equal and opposite to the original constant, cancelling it out.

This is connected with a feature of the quantum world, which shows up particularly strongly in the sum-over-histories approach, and is known as the principle of least action. In everyday language, this says that a quantum system will follow the line of least resistance from one state to another. A particle moving from one place to another, for example, will find it much easier to move in a straight line (or, rather, a geodesic) than in some convoluted path. So straight-line (geodesic) paths have a much higher probability in the sum-over-histories. The least-action principle also means that many physical features of a quantum system will tend to seek out their lowest possible level, or smallest possible value, just as water flows downhill, not upwards. In the case of the cosmological constant, it could have any value at all, including zero. So, like water flowing down a hill and following the path of least resistance, given a chance the cosmological constant will shrivel away to the smallest value it is allowed to have, which is nothing. But the important caveat is 'given a chance'. This shrivelling away cannot happen if we live in an isolated Universe; it is only possible because the Universe is connected to the meta-universe through wormholes. Then, and only then, the evolution of a Universe like our own is completely dominated by histories for which the cosmological constant is zero. Without wormholes, it is a mystery why the cosmological constant is zero today; *with* wormholes, it would be a mystery if the constant had any other value.

What's more, the same calculations tell us that the other constants of nature, such as the gravitational constant itself, must have the smallest value that is permitted, because of similar feedback effects leaking in from wormholes connecting us to other universes, and allowing the principle of least action to have full rein. It is still a big step from this to being able to calculate what the actual values of those constants ought to be, but for the first time scientists have found a hint of a reason why the laws of nature should be as they are. No wonder that Coleman refers to this as 'the big fix',* and that many theorists are now puzzling over the implications of wormhole geometry. Most of that work lies far outside the scope of the present book; but there is one implication that brings us right back to my main theme. If the structure of the meta-universe really is like a froth of bubbles interconnected by wormholes, then each of

---

* Coleman has a snappy way with words, and gave his 1988 paper explaining why the cosmological constant is zero the title 'Why There Is Nothing Rather Than Something' (*Nuclear Physics*, Volume B310, pp. 643–68). His other classic 1988 paper is called 'Black Holes As Red Herrings' (*Nuclear Physics*, Volume B307, pp. 867–82).

those bubbles – each individual universe – must be closed, in the same sense that a black hole is closed, with its own spacetime bent completely around upon itself. So, on this picture, our own Universe must be closed. That means it will one day recollapse, back into a singularity. And what happens to it then depends very much on the nature of the black holes that now exist within the Universe.

An oscillating universe?

Of course, our Universe is almost certainly closed, in this sense, whether or not it is connected by wormholes to other universes. Again, the details can be found in *In Search of the Big Bang*; but the whole business of the Universe appearing out of nothing at all, as a quantum fluctuation, depends upon it being a closed, self-contained system. The notion that the entire Universe *is* a black hole may seem bizarre at first sight, especially if you are still thinking of black holes only as superdense, compact objects. But remember that the kind of super*massive* black hole that is thought to lurk at the heart of a quasar can be made out of material scarcely any more dense than ordinary water. The bigger the black hole, the lower the density you need to close off spacetime around a collection of matter. The calculation is straightforward, and it shows that to make the entire Universe closed in this way you need the equivalent of just three hydrogen atoms in every cubic metre of space.

This, of course, is an average; it doesn't matter if many billions of those atoms are packed together inside a star, provided there are enough stars dotted across the Universe to do the trick. In fact, all the bright stars in all the bright galaxies only add up to about 1 per cent of this critical density. But there is rock solid evidence for at least ten times as much matter, probably in the form of faint stars (brown dwarfs), revealed by its gravitational influence on bright matter; and there is very persuasive evidence, from analysis of the way galaxies are spread across the Universe in filamentary chains and sheets, that there is indeed ten times more matter even than that in the form of particles filling the void. This dark 90 per cent of the Universe is known as cold dark matter, and unlike stars and galaxies it really may be distributed more or less uniformly throughout space. In which case, there are possibly dozens of cold dark matter particles passing through the room in which you are sitting, helping to hold the Universe together and make it a black hole. These would not be ordinary atoms, but a different kind of stuff altogether, left over from the big bang. Many experiments designed to capture these particles are now underway, and it seems likely that cold dark matter particles will be identified before the twentieth century is out.★

In order to see what this means for the fate of the Universe, we can go back to the old idea of escape velocity – the very concept which (although not by that name) set John Michell thinking about black holes all those centuries ago. Imagine one of Michell's dark stars, with such a strong gravitational pull that nothing, not even light, can escape from its grip. If we fire a rocket, or shoot a cannonball, upward from the surface of the star, it may rise for a time, but it is inevitable that eventually it will first halt, and then plummet back down on to the surface of the star. Now imagine the whole star swelling up, perhaps as a result of a surge of energetic activity in its heart. Each individual atom in the star will behave like that rocket, or cannonball. It can move upward (or outward) from the centre of gravity of the star for a while, but it must eventually come to a halt and then fall back. *Now* imagine that the dark star is the entire Universe, and that the atoms are replaced by galaxies. As the Universe expands, the galaxies move apart from one another. But eventually the pull of gravity will bring them to a halt, and then reverse their motion, turning the expanding Universe into a collapsing Universe that shrinks back into a singularity. The analogy is not exact, but the broad picture is good enough. That, indeed, is the fate of our Universe. And the possibility that the Universe might behave in this way was one of the options that was clear from the early days of cosmology, when the solutions to Einstein's equations were being studied in the 1920s.

Ever since that time, some cosmologists have wondered if the contraction itself could be reversed as the Universe shrank back towards the singularity. Could it be possible that at some very, very dense state, but not quite at the point of infinite density, something might happen to make the Universe 'bounce' into another cycle of expansion, so that it would actually continue eternally from expansion to contraction, through bounce to expansion, and on once again to contraction? The idea (see Figure 8.4) has

---

★ There is also scope for more exotic contributions to the total mass density of the Universe, including a possible minor influence from cosmic string left over from the big bang. But cold dark matter is far and away the front-runner for the job of closing the Universe. See, for example, my book *The Omega Point* (Bantam/Corgi, London, 1988).

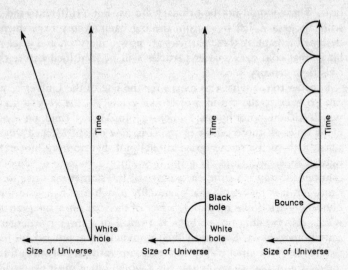

Figure 8.4 *Three possible 'histories' of time. The Universe may expand forever; it may swell up to a certain size and then shrink back into nothing; or it may undergo repeated cycles of expansion and collapse.*

obvious attractions. Not least, it resolves the perplexing puzzle of what went on 'before' the Universe began, and what will happen 'after' it has ended. But until very recently it seemed that the oscillating universe model simply could not be made to work. It conflicts with the Penrose-Hawking singularity theorems, for a start; and then there are other difficulties.

What seemed for half a century to be the definitive problem with the notion of an oscillating universe is the way entropy builds up from one cycle to the next. Entropy is the thermodynamic property that measures the amount of disorder in the Universe, and it is related to the overall temperature of the Universe. Entropy always increases, and this is a measure of the flow of time. If I were to show you a picture of a wine glass standing near the edge of a table, and another picture of the same wine glass lying in pieces on the floor next to the table, you would know which picture had been taken first – the disordered (broken) state of the wine glass must represent a later time than the ordered (unbroken) state.

Even if the Universe were to contract and start shrinking back into a singularity, it is hard to see how this would affect the flow of time and the steady increase in entropy. Although some physicists have speculated that the contracting half of the Universe might be an exact temporal mirror image of the expanding half, with time running backwards and broken wine glasses reassembling themselves, this speculation is not taken seriously by many people. It seems much more likely that entropy continues to increase in the contracting half of the cycle. Calculations along these lines were carried out in the 1930s by the physicist R. C. Tolman, and in more detail in the 1970s by David Park and P. T. Landsberg. The physical consequence of the steady build-up of entropy is that the model universes they have investigated always fall back towards the singularity harder than they emerged from it. This increased rate of collapse makes the bounce harder, so that in the next cycle the expansion starts out faster than in the previous cycle. As a result, each successive cycle expands out further from the singularity, and lasts longer, than the cycle before. Entropy rises without limit, leading to successively hotter big bang fireballs and successively longer 'life cycles' for the universe (Figure 8.5).

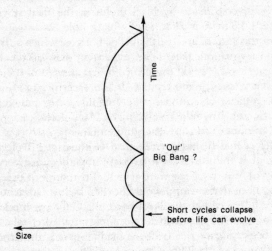

Figure 8.5 *The best buy? Our Universe may be one in a series of expansions, each one bigger than the one before, starting out from a tiny cosmic seed in the form of a quantum fluctuation of nothing at all.*

The snag with all this is that it does not, after all, resolve the puzzle of the beginning. If the Universe had already been through many such cycles, it would be very hot – much hotter than the temperature of 3 K that we measure in the background radiation we observe today, and probably too hot for life forms like us to exist. Indeed, if the thermodynamic calculations are correct, the cycle we live in can be, at most, no more than a hundred cycles after the first oscillation of the Universe that was big enough to produce stars. Of course, that might really be the case; but it does put the question of how the first tiny, short-lived version of the Universe came into being back on the agenda, and in that case most cosmologists prefer to stick with the simpler scenario of a single cycle to the Universe, starting with a unique big bang event and ending in a unique big crunch, not a bounce.

This consensus was reinforced in the 1980s, when investigations carried out by Roger Penrose suggested that it might not be possible for the Universe to be the result of a previous cycle that had bounced even once, because of a build-up of entropy far greater than anything envisaged by his predecessors. Penrose realized that nobody had taken account of the contribution to the total entropy of the Universe made by black holes in the final stages of the collapse. Like the temperature of a black hole, its entropy depends only on the surface area of the event horizon which surrounds it, and is easy to calculate. The expansion out of the big bang singularity was, we know from observations of the Universe today, very smooth and regular. It was therefore low in entropy. But the collapse of a universe like ours into a big crunch would be very different. It would involve many black holes, each possessing a large amount of entropy, merging together in a structure which is very disordered (it has been likened to squashing a piece of fruit cake so that the raisins in it overlap) and therefore very high in entropy. If 'our' big bang were actually the result of a preceding big crunch, all of that entropy would have to be lost, somehow, in the bounce. It looked like the death knell for oscillating models. Then, in one of the most dramatic developments in theoretical cosmology at the beginning of the 1990s, Canadian-based researchers ascertained that when black holes merge in the final stages of a collapsing universe the required dissipation of entropy to produce a clean new singularity really can take place.

The black hole bounce

This startling discovery emerged from an investigation of the theory of what goes on inside a realistic rotating black hole as it collapses. The study was carried out by Werner Israel, at the University of Alberta in Edmonton, Canada, and his colleagues Eric Poisson and A. E. Sikkema. Remember that the geometry of such a Kerr black hole includes two event horizons – the outer horizon, which is the last place that light can escape from the black hole, and the inner (Cauchy) horizon, which is the last place where an observer who has fallen into the hole can see light from the outside Universe. It is at the Cauchy horizon that an observer would see the entire future of the outside Universe pass instantaneously, and where the notorious blue sheet builds up. But it isn't just the blue shift of infalling electromagnetic radiation that you have to worry about.

When a realistic rotating black hole forms, it doesn't do so completely smoothly. Instead, the outer horizon settles, as Sikkema and Israel have put it, 'like a quivering soap bubble' towards the ultimate stable state represented by Kerr's solution to Einstein's equations. This quivering of the outer horizon produces ripples in spacetime – gravitational waves – that spread both outward into the Universe and inward towards the Cauchy horizon. The ones that flow outwards from the hole fade away, and are of no concern. But the ones that fall inward are blue-shifted, just like light or any other radiation that falls into the hole. But remember that energy is equivalent to mass. This inflow of blue-shifted gravitational radiation will carry energy and produce an extraordinary increase in the mass inside the black hole, increasing the core mass of a black hole which starts out with just five times as much mass as our Sun by what I remember as the Heinz factor – to 10^{57} times the mass of *the entire visible Universe*. This is a ludicrously large number. But what is even more extraordinary is that no trace of this huge mass inflation can ever show up to an observer outside the outer horizon of the black hole. The news that the mass inside the horizon has blown up in this way can only travel outwards from the Cauchy horizon in the form of gravitational radiation, at the speed of light; and therefore the information can never pass through the outer horizon and get out of the hole. From outside, an observer will still see the gravitational signature of the original five solar masses of material that collapsed to make the hole.

Because no information can get out of the black hole to affect the rest of the Universe, these calculations might seem a pointless piece of metaphysical speculation – if it were not for the likelihood that the Universe will one day recollapse into a big crunch. What will happen to these enormous masses when merging black holes overlap, and the intense gravitational fields locked away inside them can interact with one another?

To put this in perspective, in such a collapsing universe galaxies begin to overlap about a year before the big crunch. At about this time, the cosmic background radiation becomes hotter than the inside of a star, so stars break up and dissolve into a hot soup of energy and particles. Just about an hour away from the big crunch, the supermassive black holes in the hearts of galaxies begin to merge. And this changes the picture of the bounce completely from all previous models. As soon as the black holes with their enormously mass-inflated interiors merge, say Israel and his colleagues, because such intense gravitational fields are involved, the whole universe effectively collapses to the Planck scale. Instead of there still being an hour to go before reaching the singularity, the entire remainder of the collapse takes only 10^{-43} seconds (the 'Planck time'). Under these conditions, the Bekenstein-Hawking formula for the entropy of a black hole becomes meaningless, and in effect the total entropy (or entropy density) of the universe is drastically reduced. At the same time, because there is no duration shorter than 10^{-43} seconds (just as there is no distance shorter than 10^{-33} cm), the collapsing universe 10^{-43} seconds away from the singularity bounces to become an expanding universe, still 10^{-43} seconds away from the singularity, but now bursting out from the singularity, not crunching in towards it. Which rather neatly sidesteps the singularity itself, making the bounce compatible with the Penrose-Hawking singularity theorem after all.

This very nearly, but not quite, does away with the problem of the build-up of entropy from one cycle of an oscillating universe to the next. It certainly removes the problem of the huge build-up of entropy in a single bounce that Penrose was worried about. But there is still a small build-up of entropy from one cycle to the next, even though in each cycle the Universe would look as if it had emerged from a smooth singularity. Nevertheless, it does seem possible, after all, that the origins of the Universe can be traced back through successively shorter cycles to some initial tiny seed, a baby universe produced by a quantum fluctuation out of literally nothing at all.

The Reverend John Michell would, perhaps, have been uncomfortable with this notion, which seems to leave no room for God in the role of creator of the Universe. But he might well have been intrigued by the idea that the entire Universe is a black hole, which we inhabit; and that the behaviour of black holes within that Universe, merging and interacting in the late phases of a preceding big crunch, could explain how the Universe as we know it came into being. Black holes really do provide the key to understanding both the ultimate fate of the Universe and the origins of space and time.

Glossary

Accretion disk: Everything in the Universe rotates. When gas and dust in space fall on to a compact object, which might be a black hole or a neutron star, the rotation forces the infalling matter to become a swirling accretion disk of material around the central object. The accretion disk is the source of energetic outbursts of radiation from such objects.

Astronomical Unit: The average distance from the Earth to the Sun, about 150 million km (93 million miles), is defined by astronomers as the Astronomical Unit (AU) of distance.

Asymptotically: When a curving line gets closer and closer to a straight line without ever touching it, it is said to approach asymptotically. In a similar way, if you try to accelerate a moving object indefinitely it will get closer and closer to the speed of light without ever quite reaching it.

Atom: Basic component of everyday objects such as this book and your body. An atom consists of a very dense nucleus surrounded by a tenuous cloud of electrons. The radius of an atom is about 10^{-7} mm; this means that ten million atoms lined up side by side would just stretch across the width of a 'tooth' in the serrated edge of a postage stamp.

Big Bang: The outburst of matter and radiation in which the Universe was born from a singularity (or possibly from a Planck scale entity) about 15 billion years ago.

Black hole: A region of space where gravity is so strong that nothing, not even light, can escape. The gravitational field of a black hole is so powerful that light moving outwards from it is infinitely red-shifted, and loses all its energy.

Blue sheet: Light falling into a black hole experiences an infinite blue shift. This means that energy piles up in a wall around the black hole, known as a blue sheet. Any attempt to use black holes as cosmic subways, star gates or time machines will have to find a way to penetrate the blue sheet.

Blue shift: If an object is moving towards you and emits light, the light waves you see are squashed together by the motion of the object. This means they have shorter wavelengths. Because blue light has a shorter wavelength than red light, this is called a blue shift. If the Universe, which is expanding at present, ever begins to contract, a similar effect will shorten the wavelength of light from distant objects as spect itself shrinks while the light is *en route* to us. Light falling down in a gravitational field is also blue-shifted.

Closed timelike loop (CTL): A journey through space and time that returns to its starting point in *both* space *and* time, and therefore must involve travelling backwards in time for part of the journey. CTL are *not* forbidden by the laws of physics.

Cold dark matter (CDM): The large-scale dynamics of the Universe (the way galaxies move) reveals the gravitational influence of large amounts of matter that does not shine by its own light. This is referred to generically as cold dark matter; there are different theories about what the matter might be.

Cosmic censorship: The idea that there ought to be a law of nature such that every singularity must be surrounded by an event horizon so that it can never be seen from the outside Universe. Probably wrong.

Cosmic string: Thin loops of ultradense energy, far narrower than the nucleus of an atom but stretching across vast distances, left over from the Big Bang and possibly acting as the gravitational 'seeds' on which galaxies grew.

Cosmological constant: A number Einstein put into his equations of the general theory of relativity to make them match his preconceived assumption that the Universe does not expand. When observers found that the Universe does expand, the constant had no *raison d'être*; but it is still used by theorists to give more variety to their cosmological models.

Cosmological model: A set of mathematical equations that describe the evolution of the Universe. Different sets of equations (different models) are used to test theories about the origin and evolution of the universe, and make predictions (such as whether or not the model universe is expanding) which are compared with observations of the real Universe.

Degenerate stars: White dwarfs and neutron stars.

Einstein–Rosen bridge: Wormhole (Q.V.).

Electron: Particle carrying one unit of negative electric charge, found in the outer part of an atom. Each electron has a mass of 9×10^{-31} kilos (a decimal point followed by 30 zeroes and a 9 kg). Unlike neutrons, electrons are fundamental particles (members of the lepton family) and are not made up of anything else.

Entropy: A measure of information. As things wear out, entropy increases and the amount of information decreases. A glass of water with an ice cube in it has more information and less entropy than the same glass of water after the ice cube has melted. The steady increase in entropy in the Universe at large is a fundamental measure of the flow of time.

Equation of state: Equation that describes how properties such as pressure, density and temperature are related. The equation of state of a white dwarf star, for example, would allow you to calculate how the size of the star will change if its mass is increased.

Ergosphere: The region of space close to a rotating black hole from which it is possible, in principle, to extract energy.

Escape velocity: The speed at which an object such as a stone has to be thrown vertically upwards from the surface of another object such as a planet in order to escape from its gravitational pull. The escape velocity from a black hole is bigger than the speed of light.

Euclidean geometry: The geometry we learned in school, where the angles of a triangle add up to 180° and parallel lines stay the same distance apart all the way to infinity. The rules of Euclidean geometry only apply accurately to flat surfaces.

Event horizon: The surface around a black hole surrounding the region from which nothing can escape; *see* Schwarzschild horizon.

Fifth force: There are four forces of nature known to science. These are gravity, electromagnetism, and the strong and weak nuclear forces. There was a flurry of excitement in the 1980s concerning claims of the possible discovery of a 'fifth force'; careful tests now suggest, however, that those claims were mistaken.

Fission: The process by which a massive atomic nucleus fragments, releasing energy. Used in all nuclear power stations to date.

Fusion: The process by which light atomic nuclei combine to make heavier nuclei, releasing energy. The power source of all stars, including our Sun.

Galaxy: A swarm of stars held together by gravity, like our own Milky Way. A typical galaxy may contain a hundred billion stars like our Sun.

Geodesic: The shortest distance between two points. On a flat surface, geodesics are straight lines.

Gravitational constant (G): Any two objects with masses M and m attract each other with a force (gravity) that is equal to the two masses multiplied together, divided by the square of the distance between them, all multiplied by G. Isaac Newton was the first person to realize this.

Gravitational radius: The radius of the surface around a black hole (the Schwarzschild horizon) from which nothing can escape.

Hawking evaporation: The way in which a black hole radiates energy as a result of quantum effects.

Hawking process: Hawking evaporation.

Hawking radiation: The radiation emitted by a black hole evaporating by the Hawking process.

Hot dark matter: A rival to the cold dark matter theory to explain the nature of the dark stuff which holds the Universe together.

Inflation: A cosmological model describing the very rapid (exponential) expansion of the Universe when it was much less than one second old.

Inverse beta decay: For historical reasons, electrons are also known as beta rays. When a neutron in the nucleus of an atom emits an electron, and itself turns into a proton, it is said to experience beta decay. A neutron star forms when the pressure inside a star becomes so intense that electrons are forced back into protons to make neutrons – *inverse* beta decay.

Kerr black hole: A rotating black hole, which always has an ergosphere. Named after New Zealander Roy Kerr.

Light cone: The region of spacetime embraced by lines representing light rays in a Minkowski diagram. Events at a point in spacetime can only be influenced by events that occur in that point's own past light cone, and can only have an influence on events that lie in its own future light cone.

Meta-universe: The Universe is everything we can ever have any direct knowledge of. The meta-universe is everything beyond the Universe.

Minkowski diagram: A representation of the three dimensions of space and one of time as two-dimensional graphs, developed by the Lithuanian Hermann Minkowski.

Neutrino: Electrically neutral particle either with no mass at all or a very tiny mass (depending on which theory is correct), produced in some nuclear reactions (including inverse beta decay). Neutrinos are extremely reluctant to interact with everyday forms of matter and pass through the Earth more easily than machine gun bullets pass through a bank of fog.

Neutron: Electrically neutral particle with roughly the same mass as a proton, found in the nucleus of an atom.

Neutron core: Inverse beta decay may cause the formation of a neutron core in the centre of a degenerate white dwarf star.

Neutron star: A very dense, old star composed entirely of neutrons. A neutron star is in effect a single atomic nucleus containing about as much mass as our Sun in a sphere with the volume of Mount Everest.

Non-Euclidean geometry: The geometry of curved surfaces and curved space, where, for example, the angles of a triangle do not add up to 180°.

Nucleon: Collective term for protons and neutrons. Nucleons are made of quarks.

Nucleus: Central part of an atom; a ball of neutrons and protons held together by the strong nuclear force. A nucleus is about 10^{-12} mm across, 100,000 times smaller than an atom.

Occultation: When the Moon or a planet passes in front of a star, as seen from Earth.

Oppenheimer-Volkoff limit: An estimate, based on the equation of state of a degenerate star, of the maximum mass such a star can have before collapsing to form a black hole. The limit is only a few times the mass of the Sun, and has been known for more than fifty years, but was not taken seriously until the 1960s.

Parallax: The way in which foreground objects seem to move past background objects when you move your head. Used on the grand scale by astronomers, who measure the distances to the nearer stars by making observations six months apart (when the Earth is at opposite ends of its orbit around the Sun).

Photoelectric effect: The process whereby a 'particle' of light (a photon) can knock an electron out of a metal surface.

Photon: *See* Quantum.

Planck scale: Space and time may not be continuous, but rather 'quantized' so that there is a smallest length that can possibly exist and a shortest time that has any meaning. The 'Planck time' is about 10^{-43} of a second, the 'Planck length' about 2×10^{-33} of a centimetre (the distance light can travel in the Planck time), and the 'Planck mass', which is the mass that would be contained within a black hole with diameter equal to the Planck length, is 2×10^{-5} of a gram. That may sound modest, but it means that the density of a Planck black hole is about 6×10^{92} (a 6 followed by 92 zeroes) grams per cubic centimetre. A proton is 10^{20} times bigger across than such a Planck black hole.

Plasma: A kind of hot gas in which electrons are stripped from their atoms leaving behind positively charged ions. Electrons and ions mingle in the plasma. A star like our Sun is mainly composed of hot plasma.

Proton: Particle carrying one unit of positive charge, found in the nucleus of an atom. Each proton has a mass about 2,000 times greater than the mass of an electron.

Pulsar: A rapidly spinning neutron star which emits pulses of radio noise (and in some cases light and X-rays) from the magnetic field embedded in the star which spins around with it.

Quantum: The smallest bit of something that can possibly exist. For example, light energy comes in quanta which are known as photons, and can be thought of as particles of light. You cannot have an amount of light more than nothing but less than one photon.

Quantum mechanics: The set of mathematical equations that describe the behaviour of very small objects, on the scale of atoms and smaller, and radiation.

Quark: A basic building block of matter, thought to be impossible to break down into anything else. Quarks come in several varieties; protons and neutrons are each made up of three quarks in specific combinations.

Quasar: The energetic core of an active galaxy, visible far across the Universe because of its intense radiation of energy. Light from distant quasars is highly red-shifted compared with the light from relatively nearby galaxies. The energy radiated by quasars probably comes from an accretion disk around a supermassive black hole.

Recession velocity: The speed with which something is moving away from something else. The term is sometimes applied to galaxies and quasars, even though they are not moving away through space, but are getting farther apart because the space between them is expanding.

Red shift: Light waves from a distant object in the expanding Universe are stretched on the way to us, because empty space expands while they are *en route*. Red light has a longer wavelength than blue light, so this is called a red shift. A similar effect occurs for objects moving at high speed *through* space, where the motion of the object away from us stretches the light waves that we see as it is emitting them. Light climbing out of a gravitational field is also red shifted.

Schwarzschild horizon: The 'surface' of a black hole, named after the mathematician Karl Schwarzschild.

Schwarzschild radius: Gravitational radius (Q.V.).

Singularity: A point of infinite density and curvature of spacetime where the laws of physics no longer apply. Every black hole contains a singularity; the Universe may have been born out of a singularity.

Solar wind: A stream of particles moving outward through the Solar System from the Sun.

Spacelike interval: If it is possible to travel between two points in spacetime without moving faster than the speed of light, they are separated by a spacelike interval.

Spacetime: Einstein's special theory of relativity led to the realization that space and time can be described geometrically as different facets of a four-dimensional whole, spacetime. Einstein's general theory of relativity explained gravity as an effect caused by the curvature of spacetime.

Spectral lines: Bright or dark stripes in the spectrum of coloured light produced by passing white light through a prism to make a 'rainbow' pattern. Each line corresponds to an influence of one particular type of atom. By measuring the positions of lines in the spectra of distant galaxies and quasars, astronomers determine the red shift caused by the expansion of the Universe.

Star gate: Term used in science fiction for the entrance to a wormhole.

Supergiant: A very large star.

Supernova: The ultimate explosion of a very massive star at the end of its life. A supernova may briefly shine as brightly as a whole galaxy of 100 billion stars; it leaves behind a neutron star or a black hole.

Tachyon: Einstein's theory of relativity tells us that no object that has a velocity less than that of light can ever be accelerated to travel faster than light. But the equations also say that in principle objects which *always* travel faster than light, and can never be slowed down below the speed of light, might exist. They would also travel backwards in time. Nobody has ever found unequivocal proof that such particles exist, but they have been given the name tachyons ready for when (and if) they turn up. Everyday slower-than-light particles are occasionally called 'tardons'.

Timelike interval: If it is not possible to travel between two points in spacetime without moving faster than the speed of light, they are separated by a timelike interval.

White dwarf: An old star in which nuclear reactions no longer keep the core hot and roughly as much matter as is contained in our Sun has collapsed into a cooling ball about as big as the Earth.

White hole: The hypothetical counterpart to a black hole. In a black hole, matter collapses inward to a singularity; in a white hole, matter pours out from a singularity. There are some similarities between the Big Bang and a white hole.

World line: Line in a Minkowski diagram representing the life history of a particle through spacetime.

Wormhole: A tunnel through spacetime linking one black hole to another one somewhere else and somewhen else.

Bibliography

If you want to follow up some of the ideas discussed in this book, the following list will give you a start. But avoid the books marked with an asterisk if you are scared by equations.

Gregory Benford, *Timescape*, Pocket Books, New York, 1980.
Superb science fiction, from a renowned physicist, involving the many-worlds interpretation of quantum physics and tachyons – particles that travel backwards in time.

Subrahmanyan Chandrasekhar, *Eddington*, Cambridge University Press, 1983.
A brief but delightful memoir of the man described as 'the most distinguished astrophysicist of his time'. Especially interesting, in the context of black holes, for the insight into how Eddington opposed Chandrasekhar's suggestion, back in the 1930s, that nothing could prevent massive stars from collapsing to a point.

Paul Davies (editor), *The New Physics*, Cambridge University Press, 1989.
Not technical enough to merit an asterisk, but by no means a light read. Plenty of meat to get your teeth into if you want to know what is going on at the frontiers of physics today, with excellent chapters from Clifford Will on general relativity and from Malcolm Longair on astrophysics.

Arthur Eddington, *The Internal Constitution of the Stars*, Cambridge University Press, 1926.
A classic book by the pioneer of astrophysics, written just before the quantum revolution was completed; shows just how puzzled scientists were by the mystery of dense stars in the mid-1920s. Strictly speaking, this is a technical book, still used by students today – but Eddington, who among other things was a great popularizer of science, expresses himself so clearly that a warning asterisk is scarcely justified.

George Greenstein, *Frozen Star*, Freundlich, New York, 1984.
Black holes, pulsars and neutron stars described by an astronomer who has been involved in their investigation.

John Gribbin, *In Search of Schrödinger's Cat*, Bantam, New York, and Black Swan, London, 1984.
Details about how atoms work, the creation of virtual particles, and the 'many-worlds' idea.

John Gribbin, *In Search of the Big Bang*, Bantam, New York, and Black Swan, London, 1986.
My version of the story of the Universe. I'm told that there is too much detail about new theories in the last section – but it is the middle section that deals with the cosmological implications of general relativity, and I've had no complaints about that.

John Gribbin, *Blinded by the Light*, Harmony, New York, and Bantam, London, 1991.
Includes details of the structures of stars like the Sun, and how astrophysicists work out what a star is like inside.

John Gribbin and Martin Rees, *Cosmic Coincidences*, Bantam, New York, and Black Swan, London, 1990.
One of the books I most enjoyed writing – none of my own wild ideas, but presenting a view of the Universe and humankind's place in it from one of the world's leading cosmologists, Martin Rees, of the University of Cambridge.

*Stephen Hawking and Werner Israel (editors), *300 Years of Gravitation*, Cambridge University Press, 1987.
The book of the symposium held in Cambridge to mark the three-hundredth anniversary of the publication of Newton's *Principia*. Some contributions are full of hairy equations, some are readable (John Faulkner's historical insight, mentioned in Chapter One above, is not included, but only because he came up with it too late for the book's deadline, not because the editors were trying to censor this demythologizing of Newton!). Worth dipping into in a library.

Nick Herbert, *Faster Than Light*, Plume, New York, 1988.
An amazing collection of wild ideas, all of them based on extrapolation of sober scientific fact, on the theme of signals that travel faster than light and backwards in time – and all from a respectable physicist with impeccable credentials. (I borrowed the neat idea of arrows on event horizons to indicate their one-way nature from this book.)

Douglas Hofstadter, *Gödel, Escher, Bach*, Basic Books, New York, 1979.
The incompleteness theorem related to art, music and the human mind. Only tangentially relevant to the present book, but well worth reading.

William J. Kaufmann, III, *The Cosmic Frontiers of General Relativity*, Little, Brown & Co., Boston, 1977.
One of my most well-thumbed books, which explains Einstein's theory in clear language and provides an extensive overview of black holes, including drawings showing what it would be like to travel through a wormhole into another universe.

*Kenneth Lang and Owen Gingerich (editors), *A Source Book in Astronomy and Astrophysics, 1900–1975*, Harvard University Press, 1979.
An amazing collection which reprints the key parts of key scientific papers in the development of our twentieth-century understanding of the Universe. Includes Fritz Zwicky's original suggestion that neutron stars might exist, Karl Schwarzschild's mathematical description of a black hole, and much more besides, all with informative commentaries setting them in context in clear language. Fascinating to dip into, even for non-scientists, if you can find it in a library.

*Charles Misner, Kip Thorne, and John Wheeler, *Gravitation*, W. H. Freeman, San Francisco, 1973.
The standard handbook for students at graduate level. Not for the scientifically nervous, although it does include some sections presenting the basic physics in relatively simple language.

Ward Moore, *Bring the Jubilee*, Avon, New York, 1976.
Science fiction in which the South won the American Civil War – or did it? Classic exposition of the parallel universes idea.

*Abraham Pais, *Subtle is the Lord*, Oxford University Press, London, 1982.
A scientific biography that pulls no mathematical punches but has a wealth of information about the life and times of Albert Einstein.

Barry Parker, *Einstein's Dream*, Plenum, New York, 1986.
A highly readable account of Einstein's work, including some discussion of black holes, but focusing on the search for a 'theory of everything'.

Julian Schwinger, *Einstein's Legacy*, W. H. Freeman/Scientific American, New York, 1986.
One of the best books in the *Scientific American Library* series, especially good on non-Euclidean geometry and curved spacetime. It does contain some equations, but they are not the frightening kind and are more than compensated for by the excellent illustrations.

Walter Sullivan, *Black Holes*, Anchor Press/Doubleday, New York, 1979.
An enjoyable, journalistic account, built around the discovery of X-ray stars in the 1960s and 1970s, with nice illustrations.

H. G. Wells, *The Time Machine*, 1895.
The classic story, originally published in 1895 and using time as a fourth dimension ten years before Einstein's special theory.

John Archibald Wheeler, *A Journey into Gravity and Spacetime*, W. H. Freeman/Scientific American, New York, 1990.
A slightly disappointing addition to the *Scientific American Library* series, from one of the world's leading authorities on general relativity. Some nice analogies, and the usual clear illustrations that you expect from the series, but not always easy to read. Worth the effort, but not as accessible as Julian Schwinger's book.

Clifford Will, *Was Einstein Right?*, Basic Books, New York, 1986.
Simply the best guide to general relativity for the lay person.

Index

n after the page reference denotes footnotes.

Aberration, 18–19, *19*
Abramowicz, Marek, 155–7
Acceleration, 9–10, 28, 50–2
Accretion disk, 246
Adams, Walter, 68–9, 70
Aerobee rocket, 113, 114, 117
Alembert, Jean d', 23
American Association for the
    Advancement of Science, 137
American Astronomical Society, 107
American Journal of Physics, 192n, 196
American Scientist, 137
Anderson, Wilhelm, 72
Antigravity, 174, 187, 188–93, 234
Arecibo, 98
Aristotle, 3, 8
Astronomical Unit, 246
Astrophysical Journal, 75
Atom, 246

Baade, Walter, 80, 96, 104, 114
Baby universe, 232–3
Bardeen, James, 147, 149
Bekenstein, Jacob, 148–9, 151
Bekenstein-Hawking formula, 244
Bell, Jocelyn, 90–5, 96, 98, 99
Benford, Gregory, 211
Bessel, Friedrich, 66
Beta decay, 79
Big Bang theory, 102, 133, 153, 184–5,
    232, 246; and singularities, 143, 144,

178, 235; *see also In Search of the Big
    Bang*
Billiard ball paradox, 224–8
Binary X-ray sources, 115–16, 119, 120,
    123, 186
Birmingham, University of, 30
Black holes, *passim*; 1, 30, 38, 58, 61,
    85, 100–16, 120–4, 179–82, 247;
    bending spacetime, 28; and dense
    stars, 76; electrically charged, 169–73,
    175, 184–5; and energy, 109–112,
    137–9; exploding, 143, 152–3; and
    neutron stars, 81–2, 99; non-rotating,
    134, 135, 146–7; rotating, 134, 135,
    137–9, 146–7, 172–7, 179, 183;
    singularities in, 127–8, *133*, 133, 135,
    139–43; temperature, 148–9, 152, 242;
    and time, 159–60; universe in, 42,
    238, 248; virtual pairs, 151–2; *see also*
    dark stars; Schwarzchild; wormholes
Blandford, Roger, 120, 121–2, 123
Blinded by the Light (Gribbin), 65n, 71n
Blue sheet, 182–3, 185, 247
Blue shift, 178–83, 247
Blue star, 106
Bohr, Niels, 74, 125
Bolton, Tom, 119
Bolyai, János, 31, 37–8
Bolyai, Wolfgang, 37
Bouguer, Pierre, 12, 13
Bouncing universe, 239–44

Bradley, James, 18–20, 20n
Brahe, Tycho, 7, 18
Brown dwarfs, 238
Burnell, Jocelyn see Bell, Jocelyn

Caen University, 23
Calculus, differential, 3, 8, 10, 56
California Institute of Technology, 82, 106, 144, 162, 178, 221
Cambridge, 2–3, 4, 43, 71, 75, 82; Cavendish Laboratory, 13, 78, 90–5; Chandrasekhar at, 73–4; Institute of Theoretical Astronomy, 91; John Michell at, 21, 22; Radio Astronomy Observatory, 90, 103
Cambridge catalogue, 105, 106
Carter, Brandon, 147, 149, 216
Cartesian coordinates, 34–6, 45–6, 47
Casimir, Hendrik, 188–90
Casimir effect, 188–91
Cassini, Giovanni (Jean), 16–18
Cauchy horizon, 183, 243
Causality, 199, 203, 216, 223–9
Cavendish, Henry, 12–14, 21, 22
Cavendish Laboratory, 13, 78, 90–5
Cayenne, 17
Centrifugal force, 154–7, 212, 220
Chadwick, James, 78
Chandrasekhar, Subrahmanyan, 73–8, 82, 95, 126, 153
Cherenkov, Pavel, 210
Cherenkov radiation, 210
Christina of Sweden, Queen, 34
Christodoulou, Demetrios, 138, 146
Clark, Alvin, 67
Clay, Roger, 209
Clifford, William, 42–3, 56
Cold dark matter, 238–9, 247
Coleman, Sidney, 234, 237
Colour, 65, 69–70, 102, 107
Communication in Mathematical Physics, 147
Consortium, Soviet-American, 221–9, 231
Constant c, 210
Copernicus, 16
Cornell University, 145
Cosmic Censorship Hypothesis, 144–5, 154, 171, 177, 199, 247
Cosmic Coincidences (Gribbin and Rees), 178n, 183
Cosmic string, 194–5, 221, 235, 239n, 247
Cosmological constant, 184–5, 193, 194, 233, 234–5, 236, 247
Cosmological model, 101, 129, 183, 247
Cosmological red shift, 86
Cosmology Now, 142
Crab Nebula, 96, 117

Crouch, Philip, 209
CTL (closed timelike loop), 199–200, 211, 215, 216–17, 219–20, 247
Cygnus A, 104
Cygnus X–1, 117–20, 121, 143, 157

Davies, Paul, 144, 184–5, 187
Dark stars, 2, 16, 21, 27, 30, 41; and black holes, 110, 239; Laplace and, 24–5; Michell and, 22–3
David Dunlop Observatory, 119
Dearborn Observatory, 67
Dense stars, 64–99, 102, 104, 125–7; see also white dwarfs
Descartes, René, 33–6, 37, 45–6; The Method, 34; and Newton, 3, 5–6
de Sitter space, 184

Eardley, Douglas, 178, 181–3
Eddington, Arthur, 29, 70–1, 73, 126, 139, 153–4; and Chandrasekhar, 74, 75–7, 78; Internal Constitution of the stars, 71, 74
Einstein, Albert, passim; 1, 2, 30, 44, 49–53, 53–6; Kerr-Newman solution, 173; and light, 26–7, 28, 52, 123; and Newton, 55–6; Nobel prize, 26–7; probability, 201; and Schwarzchild, 58–9, 62–3; speed of light limit, 210; static universe, 100–2; wormholes, 164–5; see also general theory of relativity; special theory of relativity
Einstein-Rosen bridge, 163, 165, 170, 177, 248
Electromagnetic radiation, 98
Electron, 248
Entropy, 146–9, 240–4, 248
Epeminides, 211–12
Equation of state, 72, 73, 84, 95–6, 248
Equivalence, principle of, 51–2
Ergosphere, 136, 136, 138, 146, 149, 173, 213, 248
Eridani B, 68–9
Escape velocity, 20, 29–30, 60, 87, 127, 239, 248
Euclid, 32–3; see also geometry
Event horizon, 132, 134, 136, 139, 148, 248; and charged black holes, 184; inner event horizon, 170–1, 173, 175; and singularities, 140, 143–5, 155, 170–1
Exotic matter, 187–8, 192–3, 195
Exposition, 25

Faulkner, John, 5–6, 95
Feynman, Richard, 125, 204–8, 227–9, 236
Feynman diagram, 204, 204–7, 206
Fifth force, 15, 248

Flamm, Ludwig, 164
Fleming, Mrs, 69–70, 73
Fly (watched by Descartes), 34, 36, 45
Forward, Robert, 191
Four dimensional space, 40, 43, 45, 130, 200
Fowler, Ralph, 71–2, 74–5
Fowler, Willy, 109
French Academy of Sciences, 23
Fresnel, Augustin, 25–6
Friedman, Herbert, 114–15
FTL (faster than light) travel, 208–10

G (gravitational constant), 12–13, 15, 65, 249
Galileo Galilei, 2, 3, 5, 17
Gamow, George, 79–80, 83
Gauss, Karl, 31, 36–8, 39, 41–2, 59; and Einstein's calculations, 54
General theory of relativity, passim; 31, 42, 49–52, 53–6, 86–7, 123, 135; and centrifugal force, 212–15; and curved spacetime, 100; Kerr-Newman solution, 173; neutron stars, 85; Schwarzschild solution, 58–63, 85; singularities, 139, 142, 143; static universe, 100–1; time travel, 177, 197; wormholes, 185; see also gravity
Geodesics, 45, 56, 142, 156, 237, 249
Geometry, passim; 32–6, 37–40, 41–2, 236; and Einstein, 42–9, 54; Euclidean, 33, 38, 38, 43, 248; four dimensional, 40–9, 130, 200, 236; non-Euclidean, 31, 32–3, 38, 36–9, 39–40, 42, 59–63, 250; spacetime, 42–9, 53–63; see also Pythagoras
Gerrold, David, 203
Ginzburg, Vitalii, 105–6, 109
Gödel, Kurt, 211–12, 215, 216
Gödel's Incompleteness Theorem, 211–12
Gold, Tommy, 98, 104
Göttingen Observatory, 37, 59
Göttingen, University of, 37, 40, 82
Granny paradox, 199, 202–3, 212, 224
Gravitational energy, 79–80, 140n
Gravitational lens effect, 123
Gravitational radius see Schwarzschild radius
Gravitational red shift, 86–7
Gravity, passim; 1–2, 8–16, 20–1, 28, 87; and acceleration, 50–1; and dense stars, 67, 73; Einstein, 54–6; general theory of relativity, 52, 54–6, 62, 86–7, 123; inverse square law, 3, 8, 9, 12, 14–15, 32, 56, 62; and light, 20, 27, 42, 51–2; see also Newton
Grey holes, 179
Grossman, Marcel, 44, 49, 50, 53–4

Geometry of solar system, 66
Green Bank Observatory, 96, 118
Greenstein, Jesse, 108
Guth, Alan, 232–3, 235

HDE 226868, 119, 121
Halley, Edmund, 6, 8, 18
Harrison, Kent, 126, 129
Harvard, University of, 82, 234
Harvard College Observatory, 69, 119
Hawking, Stephen, 3, 138, 143, 145–6, 147; many-worlds hypothesis, 202; temperature of black holes, 149, 185; wormholes, 231, 234–6, 244; see also 300 Years of Gravitation
Hawking process, 149–54, 170, 180–1, 249
Hazard, Cyril, 107–8, 112, 114
Helmholtz, Hermann, 29
Helmstedt, University of, 37
Herbert, Nick, 182, 210
Hertzsprung-Russell diagram, 69
Hewish, Anthony, 90–5, 96
Hofstadter, Douglas, 212
Hooke, Robert, 4–6, 8
Hoop conjecture, 144–5
Hoyle, Sir Fred, 104, 109
Hughes Research laboratories, 191
Hyperspace, 161, 162, 177, 185, 195; see also wormholes

In Search of Schrödinger's Cat (Gribbin), 71n, 150n, 202, 207
In Search of the Big Bang (Gribbin), 133, 232, 238
Inflation, 184, 249
Interferometry, 104, 106, 118
Inverse square law of gravity, 3, 8, 9; doubts about, 12, 14–15; general theory of relativity, 56, 62; geometry, 32
Israel, Werner, 149, 243, 244; see also 300 Years of Gravitation

Jansky, Karl, 88
Jodrell Bank, 106
Jupiter, 9, 16–17, 20–1

Kaiser Wilhelm Institute, Berlin, 54
Kepler, Johannes, 5, 7–9, 16, 18, 67; laws, 7, 9, 17, 24
Kerr, Roy, 135
Kerr black hole, 173, 174–7, 182, 183, 186, 243, 249; and CTLs, 216
Kerr-Newman solution, 173–4
Kruskal, Martin, 129
Kruskal metric, 129

Landau, Lev, 75, 78–80, 82, 83
Landsberg, P.T., 241
Laplace, Pierre, 23–5, 27, 28–9, 41–2
Leiber, Fritz, 200
Leibnitz, Wilhelm, 8
Leonardo da Vinci, 196
Lick Observatory, 5
Light, *passim*; 21, 22–3, 24–7, 42, 53;
    bending, 28–9, 51–3, 55–6, 123, 142;
    and electromagnetism, 26, 40; Laplace
    and, 24–5; measuring time, 160;
    Newton and, 2, 3–5, 16, 20–1, 25;
    wave-particle duality, 26–7, 28; *see
    also* general theory of relativity
Lobachevsky, Nikolai Ivanovitch, 31,
    37–8
Lodge, Sir Oliver, 30
Lynden-Bell, Donald, 109

Many-worlds hypothesis, 202, 236
Matthews, Thomas, 106
Maxwell, James Clerk, 2, 13, 49;
    equations, 26, 27; and light, 44,
    157–8; electromagnetism, 40, 209
Mellor, Felicity, 184–5, 187
Meta-universe, 236–8, 249
Method, The (Descartes), 34
Michell, Reverend John, 13, 16, 21–2,
    245; and dark stars, 21, 24, 27, 28–9,
    41–2, 239
Minkowski, Hermann, 44–5, 48–9, 50,
    200, 213; metric for spacetime, 57,
    130
Minkowski, Rudolph, 104
MIT, 232
Moore, Ward, 202–3
Morris, Michael, 162–3, 187, 192, 196–7
Moss, Ian, 184–5, 187
Motion, laws of, 5, 43–4
Multi-dimensional space, 42, 45–9; *see
    also* spacetime
Murdin, Paul, 119

Nature, 108, 109, 149, 209–10; pulsars,
    94, 95n, 96, 98
Negative energy, 138
Negative pressure, 188, 192, 193, 194–5,
    235
Negative feedback, 186
Neutrino, 250
Neutron stars, 78, 83–5, 95–9, 120–1,
    126, 250; and supernovas, 80–2, 96;
    and time machines, 219–21; and X-
    rays, 114, 115–16, 118; *see also* pulsars
New Scientist, 216
Newcastle upon Tyne, University of,
    183
Newcastle group, 221
Newcomb, Simon, 68
Newman, Ezra, 173

Newton, Isaac, 1–16, 20, 23–4, 49; and
    calculus, 3, 8, 10, 56; gravitational
    theory, 1–2, 5, 12, 14–15, 24, 59, 87;
    and light, 16, 20, 21, 25, 52, 87;
    Opticks, 3, 6; and planetary motion,
    67; *Principia*, 1, 5, 6, 7–11, 12; and
    Robert Hooke, 4–6
NORDITA, 155
Nordstrøm, Gunnar, 170
North, J.D., 43
Novikov, Igor, 109, 178, 181, 221, 225,
    229; *see also* consortium
Nuclear fusion, 71, 79, 87, 106
Nuclear Physics, 237

Observatory, The, 76n
Occultation, 107–8, 112, 114, 250
Ohm's law of resistance, 13
Omega Point, The (Gribbin), 239
Oppenheimer, Robert, 82–8, 125, 126–7
Oppenheimer-Volkoff limit, 84, 119–20,
    124, 250
Opticks (Newton), 3, 6
Orbital motion, 7–8, 9–10, 12, 24, 67
Oscillating universe, 238–45
Oxford University, 29

Pacini, Franco, 98
Pais, Abraham, 49, 54
Palomar Observatory, 104
Paris Observatory, 17
Parallax effect, 66–7, 250
Parallel lines, 32–3, 37–9
Parallel postulate, 33, 37–9
Parallel worlds, 202, 211
Park, David, 241
Parkes Observatory, 108
Particle-antiparticle pairs, 188–9, 205–7
Penrose, Roger, 129, 136, *136*, 137, 242,
    244; on singularities, 141–3, 144, 154
Penrose diagram, 130–5, *165*, 165–6,
    169, 170–1
Penrose process, 137, 138–9, 146, 149,
    151
Penrose-Hawking singularity theorems,
    240, 244
*Philosophiae Naturalis Principia
    Mathematica* (Newton) 1, 5, 6, 7–11,
    12
Philosophical Magazine, 29, 30
*Philosophical Transactions of the Royal
    Society*, 22–3
Photons, 26–7, 251
Physical Review, 84n, 85, 129, 149; B
    192; D, 195, 216
Pickering, Edward, 69–70, 73
Pittsburgh, University of, 173
Planck scale, 231, 232, 235, 244
Planetary motion, 7–9, 57–8, 57, 59
Poisson, Eric, 243

Pope, Alexander, 2
Potsdam Observatory, 58
Princeton University, 125–6, 129, 148
Procyon, 67–8
Prussian Academy of Sciences, 54, 58
Pulsars, 90, 94–9, 105, 109, 120–1, 123, 186, 251
Pythagoras' theorem, 40, 45, 46n, 47, 224–5

Quantum theory, *passim*; 27, 30, 53, 70, 71–2, 75
Quantum uncertainty, 227–8, 230–2, 235
Quarks, 84–5, 251
Quasars, 91, 106–9, 116, 126, 178, 238, 251; and X-rays, 116, 122–4

Radio astronomy, 88–91, 100, 103–13
Red shift, 102–3, 104–5, 108–9, 181; cosmological, 102–3, 252; gravitational, 86–7, 102, 157, 159; quasar, 122
Redmount, Ian, 221
Rees, Martin, 109, 178n
Reissner, Heinrich, 170
Reissner-Nordstrøm geometry, 170, 171–2, 173–7, 182, 183
Richer, Jean, 17
Riemann, Bernhard, 31, 39, 40–2, 46, 49, 54
Ring singularity, 174–7
Rømer, Ole, 16–20
Röntgen, Wilhelm, 122
ROSAT, 113n, 122
Rosen, Nathan, 163, 164, 170
Royal Astronomical Society, 75–6
Royal Greenwich Observatory, 119
Royal Society, 4–5, 21, 22–3
Russell, Henry Norris, 69, 73

Saccheri, Girolamo, 33
Sagan, Carl, 162–4, 173, 182, 185, 186–7, 192, 221
Sandage, Allan, 106–7
Salpeter, Ed, 109
Schmidt, Maarten, 108
Schönberg, M., 80
Schrödinger, Erwin, 201–2
Schrödinger's cat, 201–2, 212
Schwarzschild, Karl, 58–63, 127; solution to Einstein's equations, 128, 134, 135, 163–4, 179
Schwarzschild black hole, *passim*; 60–3, 66, 133–4, 171–2; *see also* black holes
Schwarzschild horizon, 127–9, 133, 135, 252
Schwarzschild metric, 164, 179–80
Schwarzschild radius, *passim*; 60, 62–3, 86–7, 141, 145, 166, 249, 252; and energy, 109; and mass, 152; neutron

stars, 81, *81*; speed of light circle, 155; white dwarfs, 81, *81*
Schwarzschild sphere, 165–6
Schwarzschild surface, 128, 146
Schwarzschild throat, 167, *167*
Science fiction, 162–3, 185, 191, 199–201, 202–3, 216n
Scintillating radio sources, 90–4
Sco X–1, 113–14, 115–16, 117–18
Serber, Robert, 83
Shapiro, Stuart, 144–5
Sikkema, A.E., 243
Silverberg, Robert, 200n
Singularities, *passim*; 127–8, *133*, 133, 135, 139–43, 144, 252; and Big Bang, 178; and grey holes, 179; naked, 143, 145, 152–4, 170–1, 176, 217; and wormholes, 167, 169; *see also* ring singularities
Sirius, 67–8
Sirius B, 70, 72–3
61 Cygni, 66
Sky and Telescope, 107
Smith, Graham, 104
Snyder, Harold, 85–8, 125, 127
Solar wind, 89–90, 112–13, 252
Soldner, Johann von, 25, 28, 53
Solvay conference, Brussels, 126
Sommerfeld, Arnold, 74, 208–9
Space, *passim*; curvature of, 41, 42, 49; four dimensional, 40, 43, 45, 130, 200; and relativity, 50, 54–6, *55*
Spacetime, *passim*; 28, 42, 45–9, *46*, 59–63; diagram, 46, 130, 203–7; general theory of relativity, 56–8, 165; and gravity, 76, 83–4, 85; and matter, 56–8; warped, 32, 53, 54–5, *55*, 59–63, 100, *165*, 183; *see also* Penrose, Penrose diagram
Special theory of relativity, 27–8, 43, 44–5, 47–9, 49–52, 208; and Minkowski, 45–6; structure of stars, 72; and time travel, 217, 221–2
Speed of light, 16–20, 20n, 26–7, 28, 56–7; black holes, 60, 87; dark stars, 21, 22–3; electromagnetic equations, 26–7; gravity, 27; motion, 43–4; time, 158–9n
Speed of light circle, 155–6
Spindle singularities, 145
Stanford University, 213
Star gate, 195, 221, 252; *see also* wormholes
Starobinsky, Alex, 149
Static limit, 136
Static surface, 136, 138
Steward Observatory, 97
Stoner, Edmund, 72–3
Stoner-Anderson equation of state, 73

Sum-over-histories technique, 227–9, 236–7
Supernovas, 80, 96, 114, 252

Tachyons, 208–11, 253
Tartu University, Estonia, 72
Teukolsky, Saul, 145
Thermodynamics, laws of, 146, 147
Thorne, Kip, 140, 144–5, 173, 192, 202, 223; and CTLs, 221; and wormholes, 162–4, 182, 186–7, 195–7, 225, 227–9; *see also* consortium
3C 48, 106–7, 108
3C 144, 96, 117
3C 273, 108, 112, 114
3C 405, 104
300 Years of Gravitation, 29, 85, 120, 129
Time, 45, 158–61; *see also* time machines; time travel
Time machines, 159–61, 217–21
Time travel, 160, 162, 198–211, 212–29; general theory of relativity, 177; and time loops, 199–200, 203, 211
Tipler, Frank, 215, 221
Tolman, R.C., 241
Torsion balance, 13–14, 22
Tulane University, 215

Uhuru, 117, 147
Universal expansion, 101–3, 129, 143, 183–4, 212, 230, 234–6
Universe, *passim*; oscillating, 238–45; other, 177–8, *231*, 231–4, 236–8
University College, Galway, 29
University of Alberta in Edmonton, 243
University of California at Berkeley, 82
University of Leeds, 72
University of Maryland, 215
University of Texas, 135
University of Vienna, 211
US Naval Research Laboratory, 114

Vacuum energy, 188–92, 235
Vacuum-fluctuation battery, 191–2

Virtual Particles, 150–1, *151*, 189–90, 205–7
Volkoff, George, 83, 126
Visser, Matt, 193, 195–6, 221, 235

Wakano, Masami, 126, 129
Washington University in St. Louis, 193
Wave-particle duality, 26–7
Weber, Wilhelm, 40
Webster, Louise, 119
Wells, H.G., 200–1
Westerbrook Observatory, 118–19
Weyl, Hermann, 164
Wheeler, John, 125–6, 129, 137, 139–40, 144; *A Journey into Gravity and Spacetime*, 137, 148; many-worlds hypothesis, 202; and wormholes, 164
White dwarfs, 66–9, 70, 72–8, 81–2, 81, 85, 253; Chandrasekhar, 74–7, 126; gravitational red shift, 157; and pulsars, 95–6, 99; white dwarf limit, 75–8; and X-ray sources, 116
White holes, 129, 133, 174, 179, 181–2, 253
White Holes (Gribbin), 178n
Will, Clifford, 43, 144
World lines, 56–8, 171, 205–6, 207, 210, 213, 253
Wormholes, 163–9, 177–8, 182, 195, 253; artificial, 187–8; structure of Universe, 231–8; transversible, 163, 173, 184–8, 192, 194–7, 221–29

X-rays, 112–18, 157
X-ray astronomy, 112–24, 157

Yale University, 147
Young, Thomas, 25–6, 27
Yurtsever, Ulvi, 162, 187

Zel'dovich, Yakov, 109, 127, 149
Zel'dovich-Starobinsky effect, 149
Zurich, University of, 44
Zwicky, Fritz, 80, 88, 96, 114

READ MORE IN PENGUIN

READ MORE IN PENGUIN

SCIENCE AND MATHEMATICS

The Character of Physical Law   Richard P. Feynman

'Richard Feynman had both genius and highly unconventional style . . .
His contributions touched almost every corner of the subject, and have had
a deep and abiding influence over the way that physicists think' – Paul
Davies

A Mathematician Reads the Newspapers   John Allen Paulos

In this book, John Allen Paulos continues his liberating campaign against
mathematical illiteracy. 'Mathematics is all around you. And it's a great
defence against the sharks, cowboys and liars who want your vote, your
money or your life' – Ian Stewart

Bully for Brontosaurus   Stephen Jay Gould

'He fossicks through history, here and there picking up a bone, an imprint,
a fossil dropping, and, from these, tries to reconstruct the past afresh in all
its messy ambiguity. It's the droppings that provide the freshness: he's as
likely to quote from Mark Twain or Joe DiMaggio as from Lamarck or
Lavoisier' – *Guardian*

Are We Alone?   Paul Davies

Since ancient times people have been fascinated by the idea of
extraterrestrial life; today we are searching systematically for it. Paul
Davies's striking new book examines the assumptions that go into this
search and draws out the startling implications for science, religion and our
world view, should we discover that we are not alone.

The Making of the Atomic Bomb   Richard Rhodes

'Rhodes handles his rich trove of material with the skill of a master
novelist . . . his portraits of the leading figures are three-dimensional and
penetrating . . . the sheer momentum of the narrative is breathtaking . . . a
book to read and to read again' – *Guardian*

READ MORE IN PENGUIN

SCIENCE AND MATHEMATICS

Bright Air, Brilliant Fire   Gerald Edelman

'A brilliant and captivating new vision of the mind' – Oliver Sacks. 'Every page of Edelman's huge wok of a book crackles with delicious ideas, mostly from the *nouvelle cuisine* of neuroscience, but spiced with a good deal of intellectual history, with side dishes on everything from schizophrenia to embryology' – *The Times*

Games of Life   Karl Sigmund
Explorations in Ecology, Evolution and Behaviour

'A beautifully written and, considering its relative brevity, amazingly comprehensive survey of past and current thinking in "mathematical" evolution . . . Just as games are supposed to be fun, so too is *Games of Life*' – *The Times Higher Education Supplement*

The Artful Universe   John D. Barrow

In this original and thought-provoking investigation John D. Barrow illustrates some unexpected links between art and science. 'Full of good things . . . In what is probably the most novel part of the book, Barrow analyses music from a mathematical perspective . . . an excellent writer' – *New Scientist*

The Doctrine of DNA   R. C. Lewontin

'He is the most brilliant scientist I know and his work embodies, as this book displays so well, the very best in genetics, combined with a powerful political and moral vision of how science, properly interpreted and used to empower all the people, might truly help us to be free' – Stephen Jay Gould

Artificial Life   Steven Levy

'Can an engineered creation be alive? This centuries-old question is the starting point for Steven Levy's lucid book . . . *Artificial Life* is not only exhilarating reading but an all-too-rare case of a scientific popularization that breaks important new ground' – *The New York Times Book Review*

READ MORE IN PENGUIN

SCIENCE AND MATHEMATICS

About Time   Paul Davies

'With his usual clarity and flair, Davies argues that time in the twentieth century is Einstein's time and sets out on a fascinating discussion of why Einstein's can't be the last word on the subject' – *Independent on Sunday*

Insanely Great   Steven Levy

It was Apple's co-founder Steve Jobs who referred to the Mac as 'insanely great'. He was absolutely right: the machine that revolutionized the world of personal computing was and is great – yet the machinations behind its inception were nothing short of insane. 'A delightful and timely book' – *The New York Times Book Review*

Wonderful Life   Stephen Jay Gould

'He weaves together three extraordinary themes – one palaeontological, one human, one theoretical and historical – as he discusses the discovery of the Burgess Shale, with its amazing, wonderfully preserved fossils – a time-capsule of the early Cambrian seas' – *Mail on Sunday*

The *New Scientist* Guide to Chaos   Edited by Nina Hall

In this collection of incisive reports, acknowledged experts such as Ian Stewart, Robert May and Benoit Mandelbrot draw on the latest research to explain the roots of chaos in modern mathematics and physics.

Innumeracy   John Allen Paulos

'An engaging compilation of anecdotes and observations about those circumstances in which a very simple piece of mathematical insight can save an awful lot of futility' – *The Times Educational Supplement*

Consciousness Explained   Daniel C. Dennett

'Extraordinary ... Dennett outlines an alternative view of consciousness drawn partly from the world of computers and partly from the findings of neuroscience. Our brains, he argues, are more like parallel processors than the serial processors that lie at the heart of most computers in use today ... Supremely engaging and witty' – *Independent*

READ MORE IN PENGUIN

POPULAR SCIENCE

Naturalist Edward O. Wilson

'One of the finest scientific memoirs ever written, by one of the finest scientists writing today' – *Los Angeles Times*. 'There are wonderful accounts of his adventures with snakes, a gigantic ray, butterflies, flies, and, of course, ants ... provides a fascinating insight into a great mind' – *Guardian*

Eight Little Piggies Stephen Jay Gould

'Stephen Jay Gould has a talent for making the scientific, and particularly the evolutionary, interesting and striking ... Time and again in these essays ... he demonstrates the role of the randomness or waste in evolution' – *Sunday Times*. 'His essays are remarkable, and not just for the scientific insights at their centre ... Gould takes his readers on tough-minded rambles across the visible surface of things' – *Guardian*

Darwin's Dangerous Idea Daniel C. Dennett

'A surpassingly brilliant book. Where creative, it lifts the reader to new intellectual heights. Where critical, it is devastating. Dennet shows that intellectuals have been powerfully misled on evolutionary matters and his book will undo much damage' – Richard Dawkins

Relativity for the Layman James A. Coleman

Einstein's Theory of Relativity is one of the greatest achievements of twentieth-century science. In this clear and concise book, James A. Coleman provides an accessible introduction for the non-expert reader.

God and the New Physics Paul Davies

How did the world begin – and how will it end? These questions are not new; what is new, argues Paul Davies, is that we may be on the verge of answering them. 'The author is an excellent writer. He not only explains with fluent simplicity some of the profoundest questions of cosmology, but he is also well read in theology' – *Daily Telegraph*

READ MORE IN PENGUIN

POPULAR SCIENCE

At Home in the Universe   Stuart Kauffman

Stuart Kauffman brilliantly weaves together the excitement of intellectual discovery and a fertile mix of insights to give the general reader a fascinating look at this new science – the science of complexity – and at the forces for order that lie at the edge of chaos. 'Kauffman shares his discovery with us, with lucidity, wit and cogent argument, and we see his vision . . . He is a pioneer' – Roger Lewin

The Diversity of Life   Edward O. Wilson

'Magnificent . . . Wilson pursues the diversification of life from its earliest beginnings more than a billion years ago to the current crisis of man-driven extinction, always looking for connections, always hopeful' – *Independent on Sunday*

The Red Queen   Matt Ridley

'Animals and plants evolved sex to fend off parasitic infection. Now look where it has got us. Men want BMWs, power and money in order to pair-bond with women who are blonde, youthful and narrow-waisted . . . and all because the sexes have genes which endow them with such lusts . . . *The Red Queen* is a brilliant examination of the scientific debates on the hows and whys of sex and evolution; and the unending evolutionary struggle between male and female . . . a dazzling display of creativity and wit' – *Independent*

Voyage of the *Beagle*   Charles Darwin

The five-year voyage of the *Beagle* set in motion the intellectual currents that culminated in the publication of *The Origin of Species*. His journal, reprinted here in a shortened version, is vivid and immediate, showing us a naturalist making patient observations, above all in geology. The editors have provided an excellent introduction and notes for this edition, which also contains maps and appendices.

BY THE SAME AUTHOR

In the Beginning
The Birth of the Living Universe

Ripples in space collected by the COBE (Cosmic Background Explorer) satellite in 1992 clearly confirmed current ideas about the Big Bang. But why do matter and nature's fundamental forces seem specially designed to produce our kind of Universe? Some scientists see the hand of God; others call this a non-question; but Gribbin suggests a deeply satisfying new answer. Going far beyond the Gaia hypothesis that the Earth is a single living organism, he claims that galaxies may 'operate as supernova nurseries', that one universe can 'bud' from star-death and 'black hole bounce' into another, and that such 'offspring' are being steadily refined by evolution.

In Search of the Double Helix
Quantum Physics and Life

Scientists now understand the fundamental secrets of life. Quantum effects cause tiny genetic mutations, transmitted by DNA, which fuel the struggle for survival among plants and animals. *In Search of the Double Helix* explains how such processes interlock. John Gribbin gives an account of the fierce (and sometimes unscrupulous) races to determine the structure of DNA and crack the ultimate code. Today, he argues, even analysis of amino acids in the blood confirms basic Darwinian principles and reveals how astonishingly close we are to gorillas and chimpanzees. His book offers an ideal overview.

BY THE SAME AUTHOR

In Search of the Big Bang
The Life and Death of the Universe

Where do we come from? How did the Universe begin? And how will it end?

There have been extraordinary developments in the fields of cosmology and quantum physics since *In Search of the Big Bang* was first published in 1986. In this radically revised and updated edition John Gribbin explains the origin of the Universe and its subsequent evolution. Bringing the story right up to the present he shows how after many billions of years the Universe, which is continuing to expand, may one day collapse into a mirror image of the Big Bang. This exciting and cogent investigation into the theory of creation not only elucidates our beginnings but reveals how the Universe will end.

'A remarkably readable guide to the mysteries of cosmic creation'
– *Nature*

In Search of SUSY
Supersymmetry, String and the Theory of Everything

Many physicists believe that we are on the verge of developing a complete 'Theory of Everything' which embraces all the particles and forces in the universe. At its heart is the principle of supersymmetry (SUSY).

The quantum revolution turned the world upside-down, making uncertainty unavoidable and blurring the apparently obvious distinction between particles and waves. Life still looked relatively simple when matter was thought to consist of just neutrons, protons and electrons, but by the middle of the century the number of fundamental particles had already reached fifteen and was rising rapidly. Yet over the past few decades, as John Gribbin vividly demonstrates, we have been getting ever closer to a Theory of Everything based on quarks, superstrings (or membranes) and ingenious mathematical techniques which can reduce the four basic forces of nature – gravity, electromagnetism, the strong and weak nuclear forces – to a single superforce. Soon, he argues, scientists in search of SUSY may be able to produce experimental evidence, and thus a definitive answer to the deepest questions which have long plagued mankind.

READ MORE IN PENGUIN

The Matter Myth Paul Davies and John Gribbin

Recent developments at the frontiers of science are challenging our views about ourselves and the nature of the cosmos as never before.

In this sweeping survey, acclaimed science writers Paul Davies and John Gribbin examine the revolutionary transformation that is currently overtaking scientific thinking. From the weird world of quantum physics and the theory of relativity to the latest ideas about the birth of the cosmos, they find evidence for a massive paradigm shift. Theories of black holes, cosmic strings, wormholes, solitons and chaos challenge common-sense concepts of space, time and matter and demand a radically new world-view. Here is a truly fascinating advance glimpse of twenty-first-century science.

The Stuff of the Universe John Gribbin and Martin Rees
Dark Matter, Mankind and Anthropic Cosmology

In trying to make sense of our relationship with the cosmos, scientists have concluded that most of the universe is made up of so-called 'dark matter', the controlling factor in its dynamics, structure and eventual fate. In this illuminating account leading science writer John Gribbin and eminent physicist Martin Rees give us the most comprehensive and accessible treatment yet of the major theories and the latest advances in under-standing the nature of dark matter, which lead on to the monumental question of why the universe is the way it is.

'The great question of Life, the Universe and Everything ... a pleasure to read' – Tim Radford in the *Guardian*

READ MORE IN PENGUIN

Richard Feynman: A Life in Science   John Gribbin and Mary Gribbin

'One of the most influential and best-loved physicists of his generation ... This biography is both compelling and highly readable' – Michael White in the *Mail on Sunday*

Magnificently charismatic and fun-loving, Richard Feynman brought a sense of adventure to the study of science. As well as leaving his mark on nearly every aspect of modern physics, he was a hugely popular and respected teacher. His extraordinary career included wartime work on the atomic bomb at Los Alamos and a profoundly original theory of quantum mechanics, for which he won the Nobel prize. In 1986 he came to widespread public attention during the enquiry into the Challenger disaster when he proved conclusively that its cause was due to the effect of cold on the shuttle's rubber sealings.

Skilfully interweaving personal anecdotes, writings and recollections with narrative, John and Mary Gribbin reveal a man of startling originality who had an immense passion for life.

Stephen Hawking   Michael White and John Gribbin

'Few scientists become legends in their own lifetime. Stephen Hawking is one. It's good to have this well-documented and immensely readable biography to remind us that the media-hyped "mute genius in the wheelchair" is in fact a sensitive, humorous, ambitious and occasionally wilful human being' – Paul Davies in *The Times Higher Educational Supplement*